ADVANCE PRAISE

In a survival setting, the ability to access emergency power may make the difference between life and death. *Lights On* by Jeffrey Yago is the most extensive guide ever produced on battery power to give you the best chance to succeed if the grid fails.
—JOE ALTON, MD, AND AMY ALTON, ARNP, AUTHORS OF *THE SURVIVAL MEDICINE HANDBOOK* AND HOSTS OF THE WEEKLY PREPAREDNESS PODCAST *DOOM AND BLOOM*

As a longtime reader of *Backwoods Home Magazine*, I know Jeff Yago is an authority on solar and alternative power. For those who are totally dependent on the power grid *Lights On* will, hopefully, start the alarm bells ringing. Mr. Yago not only details the threats, he also offers solutions to overcome short- or long-term power loss. If you want to know how to live with power after the grid goes down, this is the book for you.
—JOHN EGAN, DIRECTOR PREPPERGROUPS.COM

America's enemies know the grid is our weakness. It is just a matter of time before the grid is attacked. *Lights On* is the first book you should read in your quest to be better protected.
—VINCENT VONDOOM, DISASTER SURVIVAL NETWORK

LIGHTS ON

LIGHTS ON

THE NON-TECHNICAL GUIDE TO BATTERY POWER WHEN THE GRID GOES DOWN

JEFFREY YAGO

Dunimis Technology Inc.
Gum Spring, Virginia

LIGHTS ON

Copyright © 2016 by Jeffrey Yago

All rights reserved. No part of this book may be reproduced in any form or by any means—whether electronic, digital, mechanical, or otherwise—without permission in writing from the publisher, except by a reviewer, who may quote brief passages in a review.

Published by Dunimis Technology Inc., P.O. Box 10, Gum Spring, Virginia 23065

Paperback ISBN: 978-1-7351317-1-9
eBook ISBN: 978-1-7351317-2-6

Library of Congress Cataloging-in-Publication Data: 2020912779

Book designed by Mark Karis

Printed in the United States of America

If you fail to plan, you plan to fail.
—BENJAMIN FRANKLIN

CONTENTS

	Foreword by Dave Duffy	xi
	Foreword by Jack Spirko	xiii
	Acknowledgments	xv
	Notice to Readers	xvi
	Introduction	1
1	Reality Check	8
2	Back to the Future—Again!	25
3	Life After Generator	32
4	Introduction to Battery Power	38
5	Understanding Solar System Components	52
6	Lighting with Battery Power	71
7	Emergency Communication with Battery Power	77

8	Computers with Battery Power	89
9	Entertainment with Battery Power	95
10	Medical Equipment with Battery Power	100
11	Portable Tools with Battery Power	105
12	Clean Water with Battery Power	110
13	Refrigeration with Battery Power	118
14	Security with Battery Power	123
15	Using Vehicles for Battery Power	134
16	RV Camping with Battery Power	138
17	Bugging Out with Battery Power	143
18	Connecting Devices to Battery Power	147
19	Cooking Without Grid Power	157
20	Washing Without Grid Power	161
21	Building Your Own Solar Power Supply	167
22	EMP Protection of Battery-Powered Devices	172
23	Closing Comments	182

Appendix

Useful Tables and Wiring Diagrams	185
Understanding Watts, Amps, and Volts	190
Sizing Solar Array Wire	193
Inverter Wire Sizing	195
References	199
Resources	201
About the Author	206
Notes	208
Index	209

FOREWORD

BY DAVE DUFFY

THIS IS AN EXTREMELY INFORMATIVE BOOK about using battery power to enable you to live a normal life in the event the electrical grid in your area goes down. I've already employed much of this advice in my own life, and I'm grateful to Jeff for putting it all together in such a compact format. The electrical grid in America is vulnerable for a variety of reasons: malicious computer hacking, lack of infrastructure maintenance, EMP, terrorist sabotage. Jeff Yago explains these risks and gives sensible solutions that anyone can employ.

He has taken potentially complicated material and served it up in digestible bites during the twenty-plus years he has written for *Backwoods Home Magazine*. His writing is one of the key reasons the magazine has had such staying power for twenty-seven years, and I regard him as one of the top five authors who has ever written for us. With a paid subscription base of 40,000 readers, additional newsstand and library circulation, and 250,000 unique visits to our website each month, his articles reach a large and diverse readership.

This book will be very popular among preppers, survivalists, and other self-reliant individuals. It has the sensible, easy-to-implement answers to both temporary and long-term grid-down situations. I expect it to become an essential part of the tool kit for those wise enough to start preparing for a sudden and perhaps long-term interruption in society's normal routine.

I consider Jeff Yago to be a highly regarded author and journalist, and I know you can trust what he says.

—DAVE DUFFY, PUBLISHER, *BACKWOODS HOME MAGAZINE*

FOREWORD

BY JACK SPIRKO

OVER THE CHRISTMAS HOLIDAYS IN 2012, we lived though one of the worst ice storms ever to hit Arkansas. Fortunately, because I used many of the backup-power techniques you will learn in this book, this two-week-long power outage was only a minor inconvenience for my family. When neighbors came to check on us, there was a steaming bowl of gravy on the stove and we had just taken the turkey out of the oven. All lights were on, and a football game was playing on the TV as I answered the door.

The storm took down huge trees and power lines, blocking most roads and highways, causing half a million people to be without power. Our homestead was located at the very end of a six-mile-long secondary road with only thirty families living along the entire length. Whenever there was a power outage, you can bet we were always last to get back online. By my being prepared, this ended up being a wonderful Christmas with my son and wife.

Over my years as host of *The Survival Podcast*, a daily Internet radio show, my goal has always been to help listeners build resiliency and self-sufficiency into their lives. Their greatest point of both concern and confusion is backup power. The other survival needs—food, water, shelter, and security—are relatively easy to understand and obtain. But when we turn to electrical power, we enter the mysterious world of electricians and solar experts. Satisfying this basic need for backup electrical power can be frightening for a layperson.

This book demystifies this world of backup electrical power. It will give you both the information and the confidence to build anything from a simple off-grid cell phone charger to a full-on home power system, and anything in between. This also reminds us that living on battery power was a way of life for many rural families living in the United States during the early 1900s. As cross-country power lines eventually made life easy by bringing electricity to even the remotest towns, many of us have lost the basic survival knowledge most homeowners, farmers, and ranchers knew not that long ago.

What would happen if you were forced to be without power for a week, a month, or even longer? Today everything revolves around electrical power. We use it to refrigerate and cook our food, pump our water, and maintain a comfortable climate in our homes. Electricity lights up the dark, and that also brings a measure of security. When you think about it, electricity is the keystone for binding all these other survival needs together. This book is your guide to placing that keystone into your own resilient lifestyle.

Jack Spirko, host of daily *The Survival Podcast*

ACKNOWLEDGMENTS

IT WOULD BE VERY SATISFYING TO say I wrote this book alone, but in reality it wrote itself. It took the understanding and patience of clients and product developers who allowed those few of us involved in the early beginnings of residential solar power in the late 1960s to learn by our mistakes, as books like this did not exist. Most of the pioneer solar work we did back then was trial and error, and many of the solar product designers I knew were still working out of their garages. These early beginnings eventually became the large solar manufacturers we see today providing an endless variety of quality solar and backup power systems.

Finally, this book could never have been written without the dedication, untiring efforts, word processing skills, and the endless editing by my wife, Sharon Seymore Yago.

Thanks to all!

NOTICE TO READERS

THE MATERIAL IN THIS BOOK IS for voluntary acceptance and use by the reader. This book is not meant to define wiring or safety standards, construction methods, or materials for any electrical design or wiring construction. Any included design guidelines, wiring suggestions, or product selection are intended to assist you in developing your own project outline and cost estimate.

All wiring, wiring devices, and electrical equipment you may install in new or existing construction may be subject to local building codes using electrical products listed by national testing agencies and must be installed according to the National Electric Code. It is always recommended to enlist the assistance of a licensed electrician if in doubt.

This material is subject to revision as further experience and product development may deem necessary in this rapidly changing field. Any product referred to by brand name or model number is for informational purposes only and is not intended to be an endorsement by the author or publisher.

INTRODUCTION

THAT'S ODD. THE HOUSE IS UNUSUALLY quiet this morning, and it's a work and school day. Must have forgotten to set the alarm again. The sun is up, so it has to be late. Looks like the clock stopped and my cell phone didn't charge. That old electric line must be down again. Hope they fix it right this time, and soon.

Nothing on the local news last night about any storms or high winds. We rarely have a utility problem this time of year. National news presented their usual list of disasters in other countries, but unnamed government sources said everything in this country is just fine, nothing to see, move along.

Kids starting to stir, and the oldest is yelling something about the shower not working and the toilet won't flush. Better get downstairs soon before they start trying to make their own breakfast. Not pretty.

In the kitchen now. Ice cream is melting and running down the refrigerator door onto the floor. The dog is happily cleaning it up. If the power is not back soon, we're going to lose everything in the freezer.

What I really need is news. What's happening out there, what areas are without power, and what's the weather forecast? Landline is out and the computer and Internet router are dead as well.

The car! Yes, the car! I should've thought of it sooner. The car radio should at least have local news on the outage and a weather forecast. Heading to the garage.

I didn't expect the remote garage door opener to work, but had to try. Tried manually, but it wouldn't budge. Things are really getting strange. The car remote doesn't work either, but the key works. Hmm. The engine won't turn over, and the lights and radio are also dead. This car is almost new. The battery should be good, and it's a warm morning, so what's going on?

Exiting the garage, I find that by now the kids are standing down by the road, waiting for a school bus that obviously is not coming today. It's time for a family meeting around the dining room table. I call them in.

The sun is streaming through the windows, which at least provides some natural room light. The kids are worried, and while I put on a good show, I'm also worried. Very worried.

We have had power outages before, but this time it's different. Nothing is working, not even the phones, Internet, and especially not the car. I'm starting to feel trapped in my own home, and I don't like it. This has been a great place to live, but it's starting to feel a lot less friendly.

Better assume this could last more than a day and get organized. The youngest is starting to look for flashlights that I know are around here somewhere. The oldest is heading for the garage to find the cooler and grab any refrigerator items still frozen, along with whatever ice has not melted. My assignment should be to drive into town and find out what's going on, but without transportation, that's miles to walk. Don't want to leave the kids home alone, but I'm reluctant to take them with me until I can find out what's going on.

The youngest just returned with two flashlights and a small portable radio. Unfortunately, the batteries are either dead or missing. I really did intend to stock up on batteries, but they always end up in a kids' game.

What I really want right now is a long, hot shower, but that's not going to happen.

INTRODUCTION

Back in the kitchen the sink is now piled high with the morning's cereal bowls. The dishwasher is also full of dirty dishes since I decided last night to wait until after breakfast before starting. If I can get the outdoor grill to work, we may have steaks tonight, but if the power is not back on by tomorrow, whatever we don't eat is going to spoil. A quick look through the cabinets tells me that if this outage lasts more than a few days, we all will be eating crackers and peanut butter.

I have been hearing recent ads for freeze-dried foods that only require adding boiling water, but not sure what that's all about. Without running water, a refrigerator, and a working stove, there is no way I can prepare a meal even if I had all the ingredients.

The oldest just scored some candles, most likely left over from the warming trays we used for last year's Christmas party. If I can find some matches, at least there will be light in the bedrooms and bathrooms tonight.

Speaking of bathrooms, what to do about the toilets? I have to figure something out first thing tomorrow, and better find a way to wash up. Never thought I would miss using my toothbrush.

I found two cans of fruit juice in the cabinet, but this will not last long. We must have water to drink soon, and lots of it. Government officials keep talking about the importance of storing bottled water for emergencies, and I really did intend to start doing this, just didn't think I would need it so soon.

The hours pass, with nothing to do but wait. Morning turns into afternoon, and afternoon into evening.

It's getting dark and I have yet to see any car headlights driving by, but the house is not close to the road and the trees may be blocking their lights. Haven't seen any sign of our neighbors either, but I think they are out of town this week. The kids are finally settling in for the night, so better make sure all the candles are out. Don't want to add burning down the house to the power problems.

Sounds like several people are walking down the road, but it's getting very late and very dark. I heard the sound of breaking glass earlier tonight, but thankfully, not close. This is not good. My first thought was to dial 911, but then I remembered all the phones are out.

Could this be you? Could this be your family's situation? Nothing specific, but something just doesn't feel right these days. You can't put your finger on it, but it's like an itch you can't scratch. Despite all the upbeat economic news, it just does not match your own reality. We now have to deal with politically correct speech, which does nothing but keep everyone from stating the obvious. Something is just wrong with this country, and you know it.

The United States is facing real problems. Our infrastructure is falling apart, the economy is going downhill fast, our military is being gutted, immigration is out of control, our industrial job base has left and will never return, our educational system is turning out drugged zombies that can't read and write, hackers are stealing our money and identities, the real unemployment rate is three times what the government is telling us, and the Federal Reserve has no idea what to do. The news media, Hollywood, and the politicians from both sides of the aisle continue to claim everything is fine, but you know everything is not fine.

Hopefully you realize that this government just does not have the resources, expertise, or leadership to feed and shelter millions of people if a real disaster strikes, and there have been multiple examples to drive this point home. Aging power systems, malicious computer hacking, decommissioned power plants that could not meet new EPA regulations, grid terrorist attacks, and the increased risk of an EMP all but guarantee future power outages will last much longer and occur more frequently.

So what can you do to ensure *your* family's safety should calamity arise?

WHERE TO START

There are all kinds of preparedness books available that cover long-term food storage, emergency medical care, vegetable gardening, raising chickens, water storage, and self-defense. I encourage everyone to learn more about these topics. Unfortunately, most emergency power advice is limited to using emergency generators or a cart-mounted inverter with small internal battery. A generator is just a boat anchor once it

INTRODUCTION

runs out of fuel, and a combination battery/inverter unit cannot power any major appliance more than a few hours before its small battery is totally drained.

There is almost no discussion about using battery-powered devices, yet they are perfect for extended power outages, and that is what this book is all about. I do not advocate spending thousands of dollars converting your entire home to solar power. However, I do discuss the many low-cost ways to use battery-powered appliances and lighting, and the multiple ways these can be kept charged during a major grid-down event.

It's more cost-effective to use appliances and lights during emergencies that are either designed to always operate on external battery power, or contain their own built-in batteries. Most nongenerator backup power systems use a small internal battery and inverter to convert battery power into normal 120-volt household electricity. This not only is less efficient than using battery-powered devices, but once the inverter's battery is drained, none of the connected AC appliances and lights will work.

With my all-battery, all-the-time approach, you will have a battery-powered substitute appliance and light fixture on standby. These can be brought out during an extended power outage instead of trying to keep your less efficient 120-volt AC appliances and lights operating from a single battery-powered inverter. This book also discusses the multiple ways these 12-volt DC devices can be recharged during an extended power outage, and why this is much safer than working with 120-volt AC electricity. You don't want to have all your emergency power eggs in one basket when a crisis strikes!

I purchased every battery-powered appliance discussed in this book from retail stores, so you should not have a problem finding these on your own, and you will not need to make any changes to your home's electrical wiring. After fifty years of experience designing every type of battery backup system imaginable for every type of client, I have learned one very important lesson: keep it simple!

This book explains exactly how you can have a working radio, television, laptop computer, cell phone, refrigerator, and power tools long after the grid and your generator have died. This advice will save you from making many expensive mistakes that I and other solar developers have made over the years while trying to determine what really works. In addition, you can start now by acquiring the products suggested in steps, one chapter at a time, while staying within your budget.

For example,

What's an easy way to have long-term emergency lighting without flashlights and a drawer full of batteries? Read chapter 6.

How can you hear the news and weather reports and communicate with neighbors during an extended power outage when landlines and cell service are down? Read chapter 7.

How can you access the Internet and keep a laptop computer working when the grid is down and you do not own a generator or it has run out of fuel? Read chapter 8.

How can you watch television and DVD movies, play video games, and listen to music CDs for months without grid or generator power? Read chapter 9.

Do you need a medical monitor or electronic device to sleep? How can you keep these operating night after night during an extended power outage? Read chapter 10.

How can you turn any water source into fresh, safe drinking water without the electric grid or a working well pump? Read chapter 12.

How do you keep foods cold and meats frozen long after the electric grid goes down? Read chapter 13.

How can you have a working security system during an extended power outage when these systems are normally connected to grid power and a working phone line? Read chapter 14.

INTRODUCTION

How do you change all your electronic devices over to rechargeable batteries, which brands are better, which types hold the charge longer, and how do you keep them charged indefinitely when the grid is down? Read chapter 4.

Putting together a bugout bag? You will need a way to charge your cell phone, flashlight, radio, and GPS device. Read chapter 17.

Want to install your own battery-based power system for a weekend cabin or emergency retreat? Read chapter 21.

Worried about EMP destroying all the electronic devices in your home and killing your car, and want a very simple way to protect these devices from harm? Read chapter 22.

In addition to these easy-to-read chapters, the appendix includes helpful sizing tables for those readers with a more technical background who are looking for specific instructions for wiring solar and battery-powered components.

This book was written with two goals in mind. I first want to help you prepare for a time when there will be no grid power. However, once in a grid-down event, I am hoping it will serve as a day-to-day guide to help you improvise, modify, or connect electrical devices to alternative power sources perhaps in ways never intended to keep them working. I have purposely kept each chapter brief and to the point, with lots of graphics to make it easier to visualize what you will need.

In the movie *The Matrix* Morpheus explained to Neo the if he took the blue pill he would stay in the dark so to speak. If he took the red pill he would understand what was really going on with the matrix. So I ask you do you want the blue pill and stay in the dark? Or are you ready to take the red pill and read on and be prepared? *The Matrix* will become visible to you in the next few pages if you will just join us.

1

REALITY CHECK

BEFORE GETTING INTO THE SPECIFIC METHODS and hardware needed to live a comfortable lifestyle on battery power long after the electric grid and your generator have failed, I would like to take you on a journey. First, we need to have a better understanding of the risks facing our electric grid today, as the reliability we have enjoyed in the past may not continue in the future.

It may be helpful to first review some history regarding the electric grid. You may be surprised to learn there was a time when hundreds of thousands of rural homes in America were totally powered by a large bank of batteries. I am convinced many of us may be forced to return to this past.

Most people do not realize the impact a true grid-down event will have on the United States, and this will not be anything like a week-long power outage after a hurricane. I define a true grid-down event as something that very unexpectedly happens to cause the complete failure of the electric grid over a large, multistate area. Restoring power to all affected areas would take months. Since by my definition a grid-down event is totally unexpected, I do not include power outages caused by storms. While large storms are capable of causing power outages over large sections of the country, they are rarely unexpected and most people in their paths have time to prepare, stock up, board up, or evacuate to safer areas. In addition, it is very unusual for a power outage from a storm to last longer than a few days. Even those states that suffered widespread flooding and power outages from Hurricane Katrina in 2005 and Hurricane Sandy in 2012 had utility power restored to all but the most remote areas in fewer than twelve days.

There are power outages, of course, that have no advance warning but do affect millions of people. These are typically caused by a downed power line, poorly maintained electrical switchgear, or occasional operator error. While these are helpful to review, I also do not include these as true grid-down events since they usually affect only a limited geographic area and the power is usually fully restored in a day or two.

The United States is now facing far more risks to the electric grid, which we depend on every minute of our daily lives. When a grid-down event does finally occur, large areas of the United States will be without power for months, not weeks. Critical segments of our society, including cell phone towers, hospitals, fire stations, police stations, and many government facilities, do have generator backup. However, these emergency power systems were designed to operate for only a few days or weeks.

A grid-down event will easily exhaust most generator fuel supplies, and they will just stop. I have inspected many megawatt-size emergency diesel generators at multiple government facilities, and each consumes a tanker truck of fuel each week when operating at full capacity. A real grid-down event will mean no fuel for vehicles, no traffic control, and highways clogged by abandoned vehicles, which will make refueling deliveries for these generators almost impossible, even if supplies are available from distant, unaffected areas.

The highest risk threatening today's electrical utility infrastructure is malicious computer hacking of the utility's automated control systems.

Every stage of power generation and distribution is controlled and monitored by highly automated computerized systems. Thousands of amps of electrical power flow at hundreds of thousands of volts from multiple generating plants through two hundred thousand miles of high-tension power lines to our homes and businesses twenty-four hours a day. This load and generating output constantly change every minute as millions of people turn lights and appliances on and off and as commercial facilities adjust heating and cooling system set points to changing occupancy and outdoor temperatures.

As a result, utility system computers must respond instantly to these large swings in power demand by smoothly adding or reducing generator output. Smaller hydroelectric dams and gas-turbine generator plants can be operational in minutes when extra capacity is needed quickly. However, large nuclear and coal-fired power plants may take days to bring online to handle projected major load increases. Cross-country utility lines and substations are the highways and interchanges that deliver all this power to us. Everything must stay in constant balance to avoid serious, perhaps catastrophic, damage to these utility lines, generators, and substations.

When automated control systems were first developed, each manufacturer had its own proprietary software and communication protocols. Control devices that operated large valves, switched circuits, regulated voltage, and reported alarms would not work with similar devices from other manufacturers. Pressure from the end users finally forced everyone to get together and develop standardized control devices and an open communication protocol for these control systems.

This supervisory control and data acquisition (SCADA) protocol quickly became the gold standard for all manufacturers. These SCADA controls are extremely reliable, and many have been working flawlessly since being installed back in the early 1990s. There is, however, an Achilles' heel: all these systems were designed to use hardwiring to carry their communications. When these SCADA controls were developed, there was no need to include any security or password protection, as this was before the development of the Internet. These control networks were not accessible from the outside world.

As the Internet became a major part in all industry communications, no longer were these isolated control devices just communicating across a factory assembly line or a power-generating facility. They now were being tied into the Internet and wireless communication systems so multiple facilities and remote equipment could be monitored and controlled from a central location. Factory assembly lines, remote electrical substations, gas compressor stations, backup generators, water treatment facilities, and sewage treatment plants are all now controlled with SCADA devices that communicate through phone lines, satellite modems, and Internet systems. However, their basic open communication architecture was never designed to limit access, which is mandatory when tied into today's public-access communication networks.

In March 2016 officials discovered that hackers had infiltrated the automated SCADA control system of a water treatment facility serving a U.S. city and modified the chemical mix going into the water supply. Their main goal was to steal account and credit card data for the 2.5 million customers. Although the government will not acknowledge which city was affected, hacking 2.5 million customer accounts would indicate a fairly large water treatment facility was involved, and it was evident there had been multiple entries over a two-month period. While rummaging through the web-accessible customer billing system, the hackers were able to gain entry into a separate computer network controlling all of the plant's automated water-treatment controls.

This SCADA system controlled multiple pumps, flow-control valves, and all automated chemical sensors and injectors. Fortunately, the hackers did not know what they were accessing once they left the accounting file area, so their control modifications were just trial and error. However, what if they were familiar with chemical systems and their goal was to silently poison or kill an entire city and not steal customer data? A modern water-treatment plant turns raw river or lake water into clean, drinkable tap water using all kinds of chemicals that can be dangerous in high concentrations. These include bulk tanks full of biocides, algaecides, disinfectants, flocculants, oxidants, ozone gas, scale inhibitors, coagulants, corrosion inhibitors, fluoride, chlorine, and pH conditioners, all automatically metered into the water as determined by the various SCADA controls.

I have visited numerous water-treatment plants, and due to the highly automated control systems, there were rarely more than two operators present for even the largest facilities. Consider what would happen if a hacker remotely shut off all chemical treatment and just let the untreated surface water go straight through to your faucet. River and lake water contain all kinds of pathogens, bacteria, parasites, and cysts, not to mention fertilizer and surface water runoff from upstream commercial farming operations.

While this may have been a water treatment facility, almost all water, gas, and electric utilities rely on the same SCADA control devices, and this shows they are vulnerable to hackers. If a specially designed computer virus infiltrated the control systems of an electric utility, it would be possible to override these SCADA control programs and redirect multiple sources of generating power into transmission lines and substations not designed to handle competing power flows. When overloaded, high-tension power lines quickly overheat, sag to the ground, and short out.

Transformers costing millions of dollars can overheat and their cores melt down. Simply shutting down a grid system by hacking is not the main concern. Maliciously lowering the grid voltage, switching between multiple power sources when out of phase, or rapidly cycling the power on and off will damage millions of home appliances and destroy large industrial motors designed to operate on a constant grid voltage and frequency.

Several years ago I was part of an engineering team asked to investigate the failed drive shaft of a huge, five-hundred-horsepower electric motor driving a large chiller-compressor cooling a major trauma hospital. The shaft was three inches in diameter and made of hardened steel, yet it looked as though it had been cleanly twisted in two as if it were a paper clip. Metallurgy testing found no shaft defects. After reviewing all possibilities, we focused on the maintenance staff. We found they had been manually switching from grid to generator power during weekly generator testing.

Normally, "in phase" monitor controls are used to make this transfer of power only when both power sources are synchronized, or in phase, before switching from one power source to the other. Being able to twist off a three-inch-diameter, hardened-steel drive shaft is an example of the widespread damage that a hacker could cause to large industrial equipment and assembly lines by just switching between multiple power sources out of phase. What if this motor had been driving a nuclear reactor's cooling pump, a high-rise elevator, or the floodgates on a dam?

An ill-intentioned hack into the automated systems that control the electric grid can create far more damage than just turning off the power, which is our least concern. Lowering the grid voltage without reducing load would generate a corresponding increase in amp flow, which in turn would cause cross-country power lines to overheat, stretch, and then short out, sparking fires when they touch the ground. Lowering the grid voltage can also damage the appliances in every home. And software hacking these utility control systems could block safety alarms from reaching system operators and override safety devices, resulting in the destruction of central transformers and switchgear.

In December 2015 a malware virus called BlackEnergy infiltrated the computers controlling Ukraine's national electric grid, shutting down all power to half of the country. While the outage lasted less than a day, it showed it is possible to remotely take over control of a utility grid system. This fairly unsophisticated software hack only erased computer files, causing substations to shut down. Imagine what a more sophisticated computer virus could do if it took over controls of the power grid, not just to shut down, but actually to lock out the operator's oversight controls.

In January 2016, hackers attacked the computers controlling Israel's entire electric grid just as cold winter temperatures had set in. The country needed maximum power for all the heating systems. Israel's Electric Authority was able to install software patches since that specific computer virus was already known and was fairly unsophisticated. However, by just using the Internet, a distant computer hacker successfully shut down the entire country's power grid for a day.

While this outage did not last long, it proved the controls for any utility or remote industrial equipment can be hijacked using a laptop computer connected to the Internet from anywhere. Imagine if this had been a coordinated attack from multiple points, sending much more harmful commands to these automated control systems.

Multiple congressional hearings and utility industry studies, including the July 2014 study titled *Securing the U.S. Electric Grid*,[1] have clearly identified the risk we face from targeted hacking of the utility's computer controls. However, to date there has been very little progress to harden these control systems that receive thousands of hacking attempts each day originating from all over the world. Sooner or later a malicious software attack will be successful, and the results will be devastating.

In reality, some of these computer control systems have likely already been penetrated, and malicious software is just lying dormant until a specific remote command is sent or a particular date is reached. Each high-voltage transformer in America is custom designed for a precise voltage and capacity, to match the specific transmission lines they will supply. There are no off-the-shelf "spare" transformers sitting in some warehouse just waiting to be tossed in a truck and rushed out to make repairs. Many of these large transformers are only manufactured overseas for each unique installation. They can take a year or more to custom build and ship.

Many older substations located in once very rural areas are now totally surrounded by major development. Just delivering a replacement transformer the size of a house and weighing hundreds of thousands of pounds would be a logistical nightmare. These transformers are so big that roads may need to be closed or humps leveled, overhead wires taken down, and bridges reinforced to handle the weight. In fact, there are only a handful of trailers in the entire United States large enough to haul these gigantic loads, let alone deliver them to where they would be needed.

Multiple transformer substations have already been subjected to physical attack and vandalism. These massive transformers and their inner cores require constant cooling by external fans and large cooling fins full of circulating oil. A San Jose, California, substation was attacked in 2013 by unknown assailants using high-powered rifles to shoot holes in the cooling fins of seventeen transformers, which in turn leaked out their coolant, causing these transformers to melt down. They were totally destroyed, and it took a month to patch enough of the equipment to get operational, but not fully back to normal.[2]

This substation was surrounded by a high-security fence and multiple security cameras. However, these marksmen purposely stayed out of camera view, leaving behind more than a hundred shell casings with no fingerprints. This was a well-organized, military-style attack by multiple shooters using high-powered rifles, yet the news media claimed it was just local teenage vandalism so everyone could go back to sleep with nothing to fear.

While it may be possible to temporarily reroute power flows around a single failed substation, if multiple substations are attacked at the same

A large transformer in transit. Photo courtesy of Intermountain Rigging & Heavyhaul, Utah.

High-voltage substation

time, we may not be so lucky. Congressional studies determined that the simultaneous destruction of six carefully selected substations could shut down all utility power to half of this country, and it may take up to a year to get all affected areas back to normal.

In August 2013 more than one hundred bolts securing the base of a tower carrying a 500,000-volt cross-country transmission line near Cabot, Arkansas, were cut, leaving just five bolts remaining. These last bolts held the tower upright long enough for the attackers to escape; then winds soon toppled the tower, which fell across the main rail line located just below. As the tower fell, it also cut additional lower power lines along the rail line, not to mention blocking all rail traffic for days.[3]

In June 2014 unknown saboteurs planted a homemade explosive under a fifty-thousand-gallon diesel fuel tank inside a power station near Nogales, Arizona. The tank was emergency backup fuel for the four generator turbines, which normally ran on natural gas.[4]

In March 2016 a New England utility crew discovered wires recently attached to a major power line that ran down the tower to multiple pipe bombs. A state bomb squad safely removed the pipe bombs; then utility crews removed the connecting trigger wires. Have you heard about any of these utility grid attacks taking place over the past three years? Makes you wonder why all of this is being kept so quiet.

Physical system attacks and computer hacking can cause significant damage to our utility infrastructure, but these are not the only grid-related risks we now face. An electromagnetic pulse (EMP) can not only cause extensive damage to the electric grid, but also destroy all unprotected electronic devices, including the computers controlling your car, home appliances, radios, televisions, Internet routers, modems, cell phones, and anything else that includes tiny microchips in their circuit boards.

An EMP does not harm humans and will not damage man-made structures. It will, however, cause significant damage to electrical and electronic equipment, and an EMP's energy can be collected by long, cross-country utility lines that will function like giant antennas to funnel this high-voltage spike down the lines into unprotected transformers and switchgear.

An EMP can be caused by natural events, such as the earth passing through a cloud of highly charged particles recently ejected from the sun's surface by sunspot activity. This typically occurs on an eleven-year cycle, with maximum risk during those periods of high sunspot activity. While such solar storms are not uncommon, the earth rarely passes directly through the path of these streams of just-ejected high-energy particles. However, such a collision did occur in 1859, causing extensive damage. I will describe this event in more detail in chapter 22.

A man-made energy weapon can produce the same damage. Some of the latest nonnuclear weapons technologies have the ability to be driven by or flown over a specific location and destroy all electronic devices and computers along their path. A large EMP is also a by-product of detonating a nuclear bomb at high altitude. If a single nuclear device were detonated 250 miles above the central portion of the United States, the resulting EMP would reach virtually every city and town from coast to coast, turning the United States back to a time before electricity and lasting years. (I will discuss how an EMP works and how to protect your electronic devices from an EMP more thoroughly in chapter 22.)

But whether an EMP event is caused by sunspots or a nuclear detonation makes little difference; our lifestyle will revert to the time of our grandparents. What *was* life like at that time?

Up until the early 1900s, candles or kerosene and gas lamps provided all lighting. Insulated wooden "iceboxes" kept food cold as long as the iceman delivered a block of ice several times each week. The milkman delivered milk, cheese, and butter each week, and like the iceman, he would just walk through the unlocked kitchen door and refill the icebox. Unlocked doors were rarely a problem.

Horses provided most of the transportation, although cities were starting to get streetcars and those newfangled "horseless carriages." In more rural areas, walking or riding a horse was still the major way to get from here to there. Roads often became impassable, with deep ruts and mud after every rain or snow since very few roads were paved.

In rural areas, obtaining water for drinking and washing required operating a hand pump connected to a nearby shallow well or rain cistern, and a trip to the bathroom entailed a long walk to an outhouse. Early rural clothes washing usually involved a hand washboard and large washtub. Some of the first mechanical clothes washing machines were hand operated, and all clothes were dried on a clothesline. Most homes were still heated by a woodstove, although some larger homes and city apartment buildings were starting to use a central coal-fired boiler or hot-air furnace located in the basement, with hot air ducted to air registers or steam piped to radiators in upper floors.

This was before electric-powered pumps and fans existed for heat distribution, so these heating systems operated on the principal that heat rises and the steam pipes or air ducts supplying each upper room provided an unobstructed path. These early forms of space heating were still labor-intensive and required constant monitoring to adjust the fires and control pressures. Central heating plants often required a full-time boiler operator to prevent steam boiler explosions, which were not uncommon at the time.

Food was cooked on a kitchen woodstove, and ingredients were mixed by hand. Unless you lived in a large city, life typically included raising chickens for meat and eggs, tending a garden and fruit trees, canning summer produce for winter meals, and trading with a neighbor for fresh milk to make butter and cheese. Many rural homeowners fattened up a hog with table scraps all summer; then they would butcher the animal and salt-cure or smoke the meat starting around Thanksgiving for the winter's supply of bacon and ham, which did not require refrigeration.

Unlike today, when shopping is almost a daily activity, a trip to town was a rare event made only when necessary to purchase large bags of flour and sugar to replenish the pantry, and perhaps a few sacks of seed for planting and feed for the livestock.

Although by the 1920s many rural homes still did not have electrical appliances, most did have a battery-powered AM radio. The family would gather around the radio each evening after supper for a few hours of news and maybe a live drama or comedy show. Radio programming in the 1930s included the live comedy shows of Jack Benny, Burns and Allen, Jimmy Durante, and husband-and-wife team "Fibber McGee" and "Molly." The bands of Benny Goodman, Glenn Miller, and Tommy Dorsey were the beginning of the American swing era, and listening to their remote live broadcasts from famous ballrooms in the "big city" was a regular family event. This was long before television and videotaping, so all radio programming was live, with sound-effects booths providing realism for the listening audience.

Even up to the late 1930s and just before World War II, more than half of the homes in this nation were still without electricity or phones. Daily life did not include cell phones, computers, laptops, iPads, e-mail, instant messaging, laser printers, fax machines, microwave ovens, food processors, ice machines, clothes dryers, refrigerators, dishwashers, video games, stereos, DVD players, televisions, video cameras, air conditioners, heat pumps, remote controls, freezers, and powered hand tools, since all of these were still years away from being invented. When a grid-down event does occur, the life most of us will experience will return to these earlier times. All modern conveniences will disappear.

Other than an occasional storm-caused power outage lasting a few days, most of us have had very little experience living without electricity. However, we can get a rough idea of what a power outage lasting only one night can do when it unexpectedly impacts a major city. Consider what happened in New York City on July 13, 1977. At 8:37 p.m., on just another hot, humid night during an extremely hot summer, a borough of New York City lost utility power due to a lightning strike hitting the main Hudson River substation.[5]

Although it was not actually storming, atmospheric conditions were right for significant lightning activity. Fifteen minutes later a second lightning strike took out the substation serving the nearby Indian Point Nuclear Plant. Thirty minutes later a third lightning strike took out the city's Sarain Brook substation, and another section of the city lost power. As the outage continued to spread to other areas of the city, Con Edison system operators tried to start the city's emergency backup generator plant, but this failed, since all of the plant's operators had left for the evening.

By 9:30 p.m. the three main high-voltage distribution lines supplying additional electrical power to the city from across the river in New Jersey became overloaded and their protective relays shut these down. At this point Con Edison system operators shut down power to the remaining areas of the city to avoid any further system damage. In less than one hour on one of the hottest and muggiest days of the year, the entire city of New York was in total darkness. All vehicle traffic was in gridlock, since every traffic light and streetlight had stopped working. All highway tunnels leading into and out of the city were closed due to the loss of all electric-powered ventilation and exhaust systems.

Since most buildings in New York City are high-rise, thousands of people were trapped in the totally dark and hot upper floors and in elevators stuck between floors. More than four thousand people were sweltered

REALITY CHECK

New York City during outage

in dark underground subway stations and tunnels when the all-electric subway trains came to a complete stop. Traffic controllers immediately diverted all planes heading into New York City, while thousands of travelers were left stranded in the dark LaGuardia and Kennedy airports for the next eight hours. By midnight, loosely formed gangs started to roam freely through the dark streets, stealing whatever they wanted and mugging anyone foolish enough to be outside or trying to walk home.

By the following day, after only twenty-five hours without electricity, more than 4,500 looters had been arrested and 550 police officers had suffered injuries. A single car dealership had fifty brand-new cars stolen. The thieves attached chains to these cars to pull security fencing off the fronts of department stores. There were 1,600 stores looted, and another 1,037 stores suffered major fire damage.

Since all radio and television stations were located in the city, their studios were dark and little reporting of the ongoing chaos reached the rest of the country until the next day. In a four-block area of downtown, every single store was stripped clean and over a third of Brooklyn had fire damage. Once things were finally under control, the property damage was estimated at $300 million, with all of this damage and theft occurring in just one single night.

When the finger-pointing started, it was decided that the failure was due to a combination of poorly maintained electrical equipment, aging and overloaded electrical systems, unusual lightning activity, poor operator response to correct the problem, improper utility staff training, and the long hot summer itself. None of that, however, could excuse the actions of thousands of individuals who decided to ignore the law as soon as the lights went out.

What will occur if a grid-down event causes a similar power outage, not just in one city but in many cities at the same time? What if the outage lasts weeks or months, not hours? If such anarchy is possible within hours with just the loss of electricity, what can we expect to happen in our cities when a real grid-down event occurs and, in addition to the loss of the electric grid, all water and sewer pumping stations go down, trash pickup stops, public transportation and all vehicle traffic come to a halt, all grocery stores and restaurants are dark and closed, and all fresh food and meat spoils without refrigeration?

Wait—there's more! During a grid-down event, all banks will close immediately, and all credit card processing and ATM transactions will cease. Without grid power, all computers, Internet routers, televisions, and radios will stop working. Cell phones, allowing people to talk and text with family and friends, will last only as long as their batteries charge, assuming cell towers have working generators.

No doubt roving gangs will steal whatever they want, destroy or burn what they can't steal, and rob anyone naive enough to leave home or apartment in search of food and water. City residents who do not evacuate as soon as the crisis hits will find it's too late. Those sitting in dark apartments and waiting days for the government to save them will learn that help is not coming. The problem will be too massive for relief agencies to handle.

As of 2016, New York City had a population of almost 8.5 million people who consume 25 million meals and 1.2 billion gallons of water each and every day. But following grid failure, cooking will be nearly impossible, and delivering so much food in easy-to-prepare packs is just not going to happen, especially if all public transportation has stopped and all roads are blocked by abandoned vehicles.

Without utility power, New York City's lift stations will not be able to pump more than 1 billion gallons of wastewater and sewage generated each day. Toilets and sinks in every business, home, and apartment building will quickly back up and stop working. At the same time this is happening to New York City, a real grid-down event could cause the same problems in Detroit, Chicago, Baltimore, Philadelphia, Boston, Los Angeles, Phoenix, Miami, Dallas, Atlanta, St. Louis, San Francisco, and Washington, D.C.

At 4:10 p.m. on August 14, 2003, the world's second-largest grid blackout hit midwestern and northeastern parts of the United States and Canada. This also occurred during a hot and muggy summer, with the outage affecting 55 million people in eight states. Some areas took up to a week to restore all power. The shutdown of more than 256 power plants and their high-voltage distribution systems was caused by a software "bug," keeping system operators at Ohio's FirstEnergy's central control room from receiving an overload alarm.[6]

Operators were not aware the Ohio grid was approaching maximum load when several transmission lines came in contact with nearby tree limbs during windy conditions. When system operators failed to respond due to the software bug blocking the alarm warnings, this caused the automated control systems at other, interconnected grid systems to shift their grid systems into protect mode. This in turn caused the cascading shutdown of 256 interconnected power plants that were suddenly being overloaded.[7]

Later review determined the initial problem in Ohio would have been easy for system operators to correct if they had received the overload alarms. What if someone hacked into the computer systems at multiple utility control centers and blocked system operators from receiving their alarms? What would be the impact to large population centers if this happened to multiple grid-control centers at the same time, and how would government relief agencies be able to respond?

There are now more than fourteen cities in the United States with populations above 4 million, and another thirty-seven cities with populations of over 1 million. You will not want to be anywhere near concentrations of high population during a major disruption to our utility and transportation infrastructure.

In 2009 over 46 percent of all power generated in the United States was from coal-fired power plants. However, this percentage dropped to 33 percent by 2016 and, under pressure from environmental groups, the EPA is forcing the closure of a large percentage of the remaining coal-fired plants due to their inability to meet newly mandated emission reductions. More than 170 major coal-fired power plants, with the combined capacity to power 44.7 million single-family homes in the United States, have recently closed or are scheduled to close as a result of these new regulations, and there are no plans for their replacement. Over 92 percent of the coal fueling these power plants is mined in the United States and not dependent on foreign suppliers. No doubt the national grid will have capacity issues for years after these plant closures, which will mean even more risk of power outages and rolling blackouts in the near future.[8]

Yes, this is scary, which is why we must find a way to maintain a reasonable lifestyle without the grid. It has been done before. In fact, up until the late 1930s, many homes in rural America had battery-powered lights and appliances years before the utility grid reached them. It's almost certain many of us will be returning to battery power when a grid-down event occurs. The next chapter will show that living on battery power is not a new concept. Hundreds of thousands of people relied on batteries to power their homes and farms until the Rural Electrification Act began the extension of all electric lines in 1936. It is my hope that this book will serve as a guide "back to the future."

2

BACK TO THE FUTURE—AGAIN!

THE POPULATION OF THE UNITED STATES in the early 1900s was approaching 92 million, yet 54 percent still lived in rural areas not served by any electric or phone lines. Most cities did have electric lights and basic laborsaving electrical appliances powered by early hydroelectric and coal-fired power plants. However, the rest of the country was still using kerosene lanterns and hand-operated well pumps, and was without indoor plumbing. Around this time a young electrical engineer named Charles F. Kettering was hired by the National Cash Register Company in Ohio to design

Rural farm powered by Delco-Light Plant

a small electric motor to power the first electrically powered cash register. Until then all cash registers and adding machines were hand operated.

With his early experience designing compact electric motors for cash registers, Kettering formed the Dayton Engineering Laboratories in 1909 to manufacture the first electric starters for automobiles. Up to this time all cars and trucks required a hand crank to start, which could be both exhausting and downright dangerous. To go with his new battery-powered starters, his company also designed and manufactured the first electrical systems for cars, which included a vastly improved ignition and lighting system. Until then all car headlights were polished reflectors with a gas flame that had to be lit before driving. Soon his electrical systems were being installed in every vehicle manufactured in America, and his small company became known simply as Delco.

1916 ad for Delco-Light plant

Charles Kettering had not forgotten his midwestern roots and his mom, who still used kerosene lanterns for light, since the Rural Electrification Program was still years away. Aware of this lack of electricity in rural areas of America, Kettering formed the Domestic Engineering Company in 1916 to manufacture his new invention, the Delco-Light plant. The initial model consisted of a custom-designed, two-horsepower gasoline engine coupled to a 32-volt DC generator section. The unit included a large flywheel and produced 850 watts of battery-charging power. The heavy flywheel and the relatively slow 1,400 rpm operating speed contributed to their extremely long life, which averaged forty years with regular service.

Kettering's systems were sold as a package that included sixteen glass cell two-volt lead-acid batteries that totaled 32 volts, along with a set of DC light fixtures. He installed the first working model in his mother's farmhouse, but every time he visited, he found she had switched back to her kerosene lanterns since she was not manually starting the generator, as he had instructed, when the battery charge became low.

Kettering knew that any series-wound electric generator would also act like a battery-powered motor if the electric flow was reversed, so his later production models included an automatic start feature. A mechanical relay monitored the current flowing from the batteries through the generator and into the farmhouse. Whenever the load exceeded 7 amps, this was enough current to cause the generator's shaft to rotate, which started the gasoline engine. This simple control could be manually adjusted to keep the generator running longer if the battery charge was low.

Delco-Light plant with batteries

Under normal conditions, all of the home's electric lights could be powered for several days before the generator automatically started. In fewer than six months, his new firm had 150 independent dealers and parts distributors spread across the country. His first production model including batteries and lights sold for $275, which was below cost, but he knew they would soon lower costs once they were in full production. Kettering added a larger, 3 kW model, which was sold to churches and businesses, as well as to multiple individuals wanting to share the cost of these larger generators to power their adjoining homes, becoming an early version of the future electric co-ops.

In 1913 the small, independent United Motors Car Company bought out Kettering's Delco Company, which in turn was then bought out by General Motors (GM) in 1918, and the entire operation was moved from Ohio to Detroit. This new division was called the General Motors Research Laboratories, and Charles Kettering stayed on in charge of all new product design. Since building codes at the time did not allow storing over one gallon of gasoline in a home, Kettering had designed an early version of his Delco-Light plant to operate on kerosene. Kerosene was already plentiful around a farm to fuel the kerosene lamps.

Battery-powered well pump

By the 1920s Kettering's Delco-Light plants were powering over 325,000 rural houses using his combination generator and battery systems.

To add to the functionality of his generator and battery system, Kettering designed battery-powered toasters, washing machines, clothes irons, ceiling fans, well pumps, and lightbulbs. In the 1900s almost all power tools, drills, grinders, and saws were belt driven from an overhead shaft, which in turn was powered by a steam engine or waterwheel. Since all of these early power tools included a wide drive pulley, Kettering designed a portable, 1/4 horsepower, battery-powered motor with adjustable tripod stand.

This motor included a wide pulley and flat leather belt, so the motor could be moved around the farm or shop and the legs positioned to allow the belt to power almost any existing power tool having a belt drive.

Although by the 1920s most of the larger cities had radio stations, the remote Western farms had to rely on an occasional newspaper for news. When these early radio stations began, Kettering would spend hours with his young son building radios out of scrap parts to allow listening to these distant stations. In 1923 Kettering designed the first commercial radio capable of receiving these very weak and distant AM radio stations. These radios were large wooden box affairs and the center

Battery-powered motor with belt drive

of attention in every living room.

Their glowing radio tubes needed lots of power, but these first radios were powered by heavy dry cell batteries, which could not be recharged. Later versions of the battery banks for his Delco-Light plant systems included a separate Delco Connection to provide battery power to the large wooden radios, which no longer needed their internal batteries. These extended-range radios were a marketing success and became known simply as Delcos.

Once these radios were in mass production, Kettering recognized the lack of refrigeration in rural homes, which did not have the city convenience of the iceman coming around to replenish the large block of ice in everyone's iceboxes. To solve this lack-of-ice problem, Kettering helped invent Freon gas in association with the DuPont Chemical Company, which he and his team used to design the first residential refrigerators. He named these the Delco Frigidaire, which operated from his 32-volt DC off-grid battery and generator systems. Models were later designed to operate

Battery-powered Delco Radio

Battery-powered refrigerator

on 120-volt AC grid power that was starting to reach city homes, and they also became an instant success. The iceman was soon out of a job thanks to the "Frigidaire."

While all this was going on with Charles Kettering back east, in 1928 a new midwestern company was being formed by Marcellus and Joseph Jacobs after they had developed a homemade wind-powered generator to power their Montana ranch. After great local success building "Jacobs Wind Energy Systems" for neighbors, in 1931 the brothers moved operations to Minneapolis, where they sold more than twenty thousand "Jacobs" wind generators through a system of dealers.

Their dealers quickly recognized they already had a ready supply of customers by selling their wind-powered generators to charge the existing 32-volt battery banks connected to the Delco-Light plant generators. Midwestern farms were in ideal locations for wind power, with plenty of wind flowing across the treeless prairies. Farmers found they rarely needed to fire up the Delco-Light plant to keep the batteries charged after adding a Jacobs wind generator. Of course, like the buggy whip, all good products eventually reach their end of life,

Ad for Delco Battery–powered washer

and this was also true of the Delco-Light plants and Jacobs wind generators.

When the federal government's "Rural Electrification Act" was passed in 1936, there were over 350,000 homes being powered by batteries charged by gasoline and wind generators. As this act slowly extended the electric grid to rural America by providing low-interest loans to start electric co-ops, all of these fully operational Delco and Jacobs DC power systems were eventually scrapped or abandoned in place, never to run again except for a few collectors and museums.

Wind-driven battery charger

Charles F. Kettering continued to develop new products for GM until retiring in 1947, having accumulated twenty-one honorary doctorate degrees. Before his death in 1958, Kettering worked with Alfred P. Sloan, president of GM when Kettering was director of product development, to start the Sloan Kettering Institute. Now called the Memorial Sloan Kettering Cancer Center, today it is a world-renowned cancer research facility.

It has been a hundred years since Kettering invented the Delco-Light plant and in essence began the off-grid battery-powered home revolution. If recent geopolitical events are any indication, I feel many rural homeowners may be forced to return to our past to power their future.

The next few chapters will review each grid-powered device you are currently using and ways to do the same thing using battery-powered devices. Unlike grid-powered devices, there are multiple ways to keep battery-powered devices charged and operating when everything else fails.

3

LIFE AFTER GENERATOR

THROUGHOUT THIS TEXT THE PRIMARY EMPHASIS will be the many ways you can use battery-powered appliances long after the grid has gone down and your generator has run out of fuel. While I do not expect anyone to be able to operate indefinitely on generator power alone, there are ways to stretch a limited supply of fuel during a normal power outage.

Generators are great until they run out of fuel.

The most common generators sold today for residential backup power are the whole-house propane-fueled generators in the 12 to 20 kW size range, and portable gasoline-fueled generators in the 4 to 8 kW size range. Regardless of which type generator you are considering, owning a generator is like adopting a new member of the family. You now have something that must be properly maintained and regularly tested, and you must always have on hand an adequate supply of fresh fuel.

Most inexperienced non-preppers will race to the builder supply warehouse the day after a major storm and buy whatever portable generator has not been sold. They then rush home, start it up, and run extension cords to the television, refrigerator, and a few lamps. Two days later, the crisis is over, and the generator is shoved into a corner of the garage to wait for the next power outage, which for most may be a year or more away.

In the meantime, the remaining gasoline in the tank will have turned to turpentine, which has rusted the gas tank and fouled the carburetor, and the ethanol blend has dissolved the plastic fuel lines and all of the rubber gaskets. What's amazing is this non-prepping homeowner is actually surprised that the generator will not start when needed again next year! Although there are fuel additives and stabilizers to extend the life of stored gasoline, they still are intended for only the few months of off-season fuel storage, not to offset a year or more of neglect.

If you choose a quality portable generator as your primary emergency backup, keep the fuel tank near empty, exercise it regularly as directed in the owner's manual, and keep multiple five-gallon containers of gasoline on hand that are numbered and rotated to the lawn mower to keep them fresh. I still recommend adding a fuel stabilizer to each of the containers, but never, ever use ethanol-blended gasoline to fuel any small engines, yard equipment, or a gasoline generator.

Regardless of advertising claims, all of my gasoline-powered lawn equipment stopped working due to dissolved fuel lines and seals after using ethanol-blended gasoline for one summer. Each of the separate service technicians at the different repair centers indicated the problem was ethanol-blend fuel, and they all said they see this type of damage every day. When preparing for emergencies, this is no time to risk a failed generator due to fuel issues. This may mean you will have to drive across town to find a gas station that still offers ethanol-free gasoline to fuel your generator, chain saw, log splitter, riding tractor, and string trimmers, but it will be worth it.

Almost all whole-house backup generators are designed to monitor the utility power and start as soon as a power outage is sensed. While this is very convenient when the power goes out in the middle of preparing a meal, and seconds later you can continue whatever you were doing on generator power, the problem with this setup is that your home may be unoccupied at the time of a power outage and the only thing needing power is the clock on the microwave. In addition, these systems are not designed for long-term, continuous operation, which will void the warranty, but it is doubtful your fuel supply will last longer than a week anyway when operating twenty-four hours per day.

If the power just went out and your generator has started, your first task should be to determine if this is a typical outage due to a storm or perhaps a local line failure that will soon be repaired. All utilities offer automated repair status reports by phone and websites that continually update repair schedules. However, if this outage is something different or just does not feel right, it may be prudent to turn off the automatic start feature and start getting out the LED lanterns, flashlights, and battery-powered radio you hopefully can find before manually shutting down the generator and associated 120-volt AC lighting and appliances.

A power outage when there is no storm in your area or in the forecast is very unusual and should ring alarm bells. The next step is to recognize that not everyone will have a generator, and if the power outage lasts longer than a few days, a generator will draw people you may not want to your location. During a typical utility outage when everything comes to a complete standstill, a residential generator running on a quiet night can be heard for a great distance, and such generators have been known to walk away if not chained down.

If you do determine that the power outage may last far longer than expected, to conserve fuel you should only start the generator manually for an hour or two in the morning while everyone is getting showered, dressed, and eating breakfast. You can then run the generator again for a few more hours during dinner and kitchen cleanup, and perhaps another half hour for television or computer use to catch up on the news.

If you keep the refrigerator and freezer doors closed and avoid constantly opening them, the limited operating hours of the generator during the morning and evening should still be enough to keep everything cold or frozen.

If you are well into a power outage lasting more than a few weeks, your concern will be security and availability of food and water. Hopefully you have already upgraded your door locks and taken other safety precautions. However, a house brightly illuminated at night when all others are dark is not safe. Several of my clients with extensive backup power systems have heavy blackout drapes for the main rooms, where the family will be gathered at night during an area power outage.

Those with portable generators can maximize fuel efficiency by charging up all battery-powered devices at the same time while also powering a much larger load or cooling down the freezer. If located in a rural area, this may also include a well pump or microwave oven. Regardless, it's important to understand that the fuel consumption of any generator is not a linear function of load.

A lightly loaded generator will use far more fuel per watt-hour produced than a generator operating closer to its design load. Do not operate a generator just to charge a cell phone or similar small load.

Charge multiple battery-powered devices while running the generator for the major power-consuming tasks, as this charging typically adds very little generator load.

Older-design whole-house generators have a separate charging output to keep the starter battery charged the same way the alternator in a car charges the car battery while you are driving.

However, many newer whole-house generators have a small starter battery that uses a "trickle" charger connected to a house circuit to keep the battery charged and is *not* charged by running the generator. This means it's possible for the starter battery to lose all starting capability during an extended power outage even if the generator has been operated recently.

This is especially true with newer generators with a diagnostics display panel, which is constantly using a small amount of battery power twenty-four hours per day even when the generator is not running. This is why manufacturers now include a grid-powered battery charger with their whole-house generators. A small solar battery maintainer can be mounted on the top of the generator enclosure to keep this battery charged regardless of grid or no grid, as long as it is not located in the shadow of the house or trees.

This is a very common problem with total off-grid solar homes having a backup generator with the digital diagnostics display, as they do not have a grid-supplied outlet to provide constant power. Several of my off-grid clients have totally disconnected the diagnostic display panels and automatic oil sump heaters in their generators to stop this constant drain on their starter batteries.

Many homeowners rely on a generator as their only source of emergency backup power. Unfortunately, backup generators are not designed to operate continuously during a true grid-down event. If you doubt me, take a look at a typical warranty for a whole-house generator and you will see it's based on a fixed number of operating hours per year, not per month. Having generator backup also assumes normal grid power will be restored in a week; then a quick trip to the local gas station or call to the local propane supplier can refill the tank.

I am sorry to point out the elephant in the room, but in a real grid-down event, it's not just your local area that will be without utility power. Major sections of this country will also be without electricity. Highways will be in total gridlock, with cars and trucks abandoned by their drivers when they run out of fuel or their electrical systems fail from an EMP event. Not only will your local gas station be closed, but most likely every gas station, store, and parts supplier for hundreds of miles will also be empty and dark. It could take months before basic utility service can be restored to all areas.

To be truly prepared for a long-term power outage, you will need a long-term solution. It's time to consider acquiring the many different battery-powered devices that are now available that can totally replace most 120-volt AC grid-powered devices during an extended power outage including 12-volt DC kitchen appliances, freezers, radios, televisions, satellite systems, medical equipment, well pumps, and power tools. The next few chapters will help you evaluate the battery-powered version of these appliances and electronic devices, plus the multiple ways to keep them charged without the electric grid or a generator.

4

INTRODUCTION TO BATTERY POWER

UNLIKE GENERATORS, WHICH PRODUCE 120- AND 240-volt AC electricity to match the power delivered to your home from the grid, battery-powered devices typically operate from a DC battery or some type of charging device that produces DC electricity from some other power source.

AC/DC POWER EXPLAINED

The terms *AC* and *DC* describe different types of current flow in a circuit. *DC* stands for *direct current*, and the electricity flows in only one direction. *AC* stands for *alternating current* since the current flow alternates its direction. If you could see the AC current flowing into your home's electrical system, it would produce a sine wave–shaped voltage curve that reverses from a peak positive to a peak negative voltage sixty times each second. Since a DC current does not reverse direction, its voltage would produce a curve that is a straight line as shown below. Almost all power supplied from central generator plants regardless of fuel used is AC because a DC voltage cannot be increased or decreased using a transformer. It's easier to generate and transport electrical power across long distances when raised to hundreds of thousands of volts before leaving the power plant. For example, by doubling the voltage, the same amount of electrical power can be transported with half the current, which allows using smaller and less expensive transmission lines. When this very high voltage power reaches the edge of your town, substations lower the voltage back down to a lower voltage, but still in excess of several thousand volts depending on which section of the grid is being served. When this power reaches your street there will be a small pole-mounted transformer, or pad-mounted transformer for underground wires, for each house or business which reduces this middle range voltage down to the 120/240-volt power supplied to your home's circuit breaker panel.

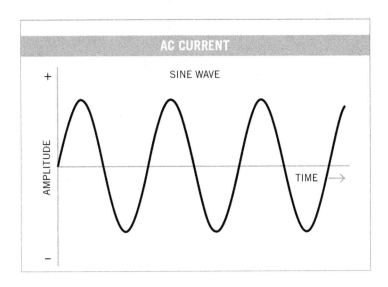

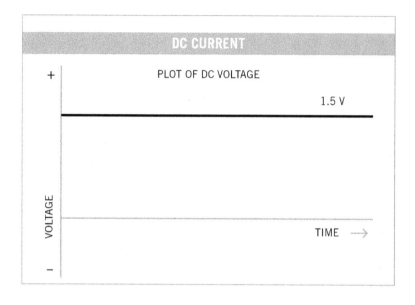

While AC electricity has many advantages for long-distance transmission and the ability to use transformers to raise and lower its voltage, the one thing AC electricity cannot do is be stored directly in a battery since all batteries are DC voltage devices. This means any electrical appliance or electronic device designed to work on AC power will not work without grid or generator power.

While it is possible to use an inverter to convert DC battery power into AC power for these loads, this is very inefficient and none of these AC appliances will work if there is a problem with the inverter or battery bank. During a major outage, it is much safer and more reliable to use individual battery-powered devices that utilize their own batteries, which can be charged using multiple backup power sources.

There are all types of battery-powered devices that can replace their 120-volt AC counterparts during an extended power outage and long after your generator has died. Battery-powered devices must also be more energy efficient than grid-powered devices to maximize battery life. This makes battery-powered devices a better choice to have during an extended power outage than trying to keep grid-powered devices operating using an emergency generator or backup inverter power. Their smaller size and portability are also ideal if you have to relocate or bug out during some type of disaster.

Until now we have not actually discussed battery technology, or which types of batteries to look for when purchasing battery-powered devices. Of course, you could just buy a fresh pack of non-rechargeable batteries every few months, just in case there will be a power outage in the near future. However, every time I have tried this approach, they either have lost their charge or started to corrode while still in the package long before I had a chance to use them.

While it is still a good idea to keep a few packs of non-rechargeable AA and C batteries around for your small flashlights and portable radios, converting over to rechargeable batteries is a better long-term solution. This is especially true for digital cameras and similar devices that seem to eat batteries. The last digital camera I owned ran on four AA batteries and could easily go through sixteen batteries in a day of project documentation. Cell phones and portable computer devices are now manufactured with very-long-life rechargeable batteries that stay in the device during charging and cannot be removed. However, most portable battery-powered flashlights, radios, walkie-talkies, digital cameras, and TV remote controls are normally furnished with disposable AAA, AA, or C batteries.

Think of all the times you needed a flashlight, only to find it buried in a kitchen drawer with dead batteries. While you could pick up a pack of flashlight batteries the next time you go to the store, this simple act will not be possible during a grid-down event. Ever go to a grocery store the day before a snowstorm was forecast? The first things to sell out will be bread, milk, bottled water, and flashlight batteries, so you can imagine how impossible it will be to find batteries during an extended power outage that could last weeks or even months for a true grid-down event. How will you power all those battery-powered devices once their disposable batteries die?

Discarded button batteries containing mercury-oxide contribute to almost 90 percent of the mercury going into today's landfills, while dry cell batteries contribute to half of all cadmium and nickel found in our landfills. It is estimated that the United States discards over 3 billion dry cell batteries each year, which is more than thirty-five batteries per family. While this may be a reasonable national average, for homes with teenagers, I think the number is far higher.

To extend the life of disposable batteries, many manufacturers are switching to heavy metals in their battery construction, which can include nickel, cadmium, lead, mercury, and acid. If discarded in landfills, these potentially toxic materials can leach into our lakes or groundwater or, if incinerated, can become harmful airborne ash. Switching to rechargeable batteries has the added benefit of reducing these damaging effects.

If you are like me, your first impression of rechargeable batteries was not good. Every time you needed to use a flashlight or digital camera having rechargeable batteries, the batteries were dead and you had to wait all day for them to recharge. Today's rechargeable batteries are different, and you really need to start using them on a regular basis in all your battery-powered devices, as they will be priceless during a grid-down event.

To make this change, you need to first purchase a minimum of two sets of rechargeable batteries for each battery-powered device, and leave one set in the device, while the other set can be charged up and ready to switch out.

Second, you need one of the new multibattery chargers, which can charge four to eight batteries at the same time, regardless of battery size or type. It makes no difference regarding the mix of battery types or sizes.

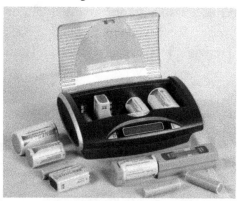
Digital-controlled fast charger

As soon as a battery is placed in the charger, a diagnostic test determines the type of rechargeable battery and its existing level of discharge. The charger then selects the best procedure to use for recharging each battery as fast as possible without risking battery damage from overheating. Some lower-cost battery chargers do not monitor temperature during charging and can damage the battery due to improper charging voltage or when a battery is left too long in the charger.

Centralize your location for all chargers and batteries.

Third, you need a central location for your battery chargers and storage for already-charged batteries waiting to be switched with the discharged batteries just removed from a device. Since you may also have a separate cell phone charger, laptop computer charger, and perhaps a charger for a Bluetooth device, keep all of these chargers together. Use a switched outlet strip to power all of these chargers, which allows turning everything off when all charging has been completed. This saves wasted energy, as most of these chargers still consume power even when not charging.

Finally, you need lots of rechargeable batteries. Many of us grew up with the older, Ni-Cd battery technology, which had "memory" problems, and if not fully discharged before recharging, the charging process would end before reaching a full charge. In addition, most of the earlier rechargeable battery technologies would lose the charge in a few days and need to be recharged again immediately before using.

COMMON HOUSEHOLD BATTERY SIZES

SIZE	DIAMETER MM (INCH)	LENGTH MM (INCH)	VOLTAGE
AAA	10.5 (0.41)	44.5 (1.75)	1.5
AA	14.5 (0.57)	50.5 (1.99)	1.5
C	26.2 (1.03)	50.0 (1.97)	1.5
D	34.2 (1.35)	61.5 (2.42)	1.5

Today's rechargeable batteries are higher quality and have a much longer life than earlier designs. The most popular rechargeable batteries sold today are nickel–metal hydride (Ni-MH). The nickel-cadmium (Ni-Cd) rechargeable battery is an older technology and is being replaced by the newer, Ni-MH batteries. The Ni-MH battery technology will store two to three times more energy than the older, Ni-Cd technology and can be recharged over one thousand times.

There is also an updated version of the Ni-MH battery technology, called LSD NiMH, which has an extremely long self-discharge (LSD). These can hold 90 percent of their charge for over a year and 75 percent of their charge for up to two years. Even without the LSD design feature, most rechargeable batteries will hold their full charge for at least nine months.

The newer, lithium-ion (Li-Ion) batteries typically found in today's cell phones, laptop computers, and portable power tools provide even more charge capacity while weighing less. However, most of these are either built into the device or are supplied with their own special charger that matches their unique shape and charging cycle requirements. Lithium-ion batteries are barred from airfreight shipping, as several recent cargo hold fires have been traced to these batteries spontaneously combusting.

Several manufacturers of larger battery-backup systems have recently switched from using older, lead-acid battery technology to the newest lithium iron phosphate (LiFePO$_4$) battery technology.

RECHARGABLE BATTERY TECHNOLOGY

TYPE	CHEMISTRY	DISADVANTAGES	ADVANTAGES
Ni-Cd	Nickel-cadmium	• older technology • high self-discharge • charging memory effect • fewer deep recharges	• very inexpensive
NiZn	Nickel-zinc	• high self-discharge • fewer deep recharges • higher cost • needs special charger	• higher voltage • good for high-demand devices • no charging memory effect
Ni-Mh	Nickel-metal hydride	• higher cost • needs special charger	• good for high-demand devices • no charging memory effect • good all-around rechargeable battery • newer battery technology
Li-Ion	Lithium-ion	• flammable electrolyte • needs special charger • charging can overheat • much higher cost	• very compact size • high energy density • good for high-demand devices
LiFePO4	Lithium-iron phosphate	• higher cost • less energy density	• good for high-demand devices • no charging memory effect • newer battery technology • much safer than Li-Ion

Although these are three times the cost, for the same amp-hour capacity they are one-third the weight and four times the cycle life. In addition, their very low, 2 percent-per-month self-discharge loss is far below many other standby battery types.

Try to standardize on only two or three battery sizes when purchasing any new electronic devices to make it easier to have freshly charged batteries ready for use. For example, AAA batteries are very small and have limited operating time even when freshly recharged. Some LED flashlights and digital cameras require AAA batteries, and some require AA batteries. The AA batteries hold substantially more charge than the smaller, AAA batteries, so the larger, AA battery should be the minimum size of rechargeable batteries you use. Larger LED lanterns and battery-powered radios typically use either C or D batteries.

You will probably find more LED lanterns and portable radios using C batteries these days than the larger D to reduce device size and weight, so the C rechargeable battery should be your second rechargeable battery size to have on hand. The third size of rechargeable battery to have is the rectangular 9-volt battery, which are typically used in handheld radios, electronic games, smoke detectors, and small electronic devices.

When I started my conversion to all rechargeable batteries, I found a wide difference in charging capacity from one battery manufacturer to another. For example, the highest-quality C-cell rechargeable batteries from Sanyo have a 6,000 milliamp-hour (mAh) rating, while some less expensive C-cell batteries are in the low 4,000 mAh range. While less expensive, this 33 percent lower charge capacity would translate into a similar drop in device operating hours.

Since these batteries have a much smaller charge capacity than the amp-hour ratings used for comparing car batteries, the smaller-scale milliamp-hour (mAh) rating is used, which is just amp-hours divided by 1,000. Any AA battery with a 3,000 mAh rating will have twice the stored power of a similar-sized battery having a 1,500 mAh rating. Unfortunately, these ratings are not always easy to identify without reading the small print, so when you are shopping for rechargeable batteries, remember: the brand that costs half as much may also have half the recharge capacity.

There are some rechargeable battery brands and models that consistently outperform all others. Professional photographers needing

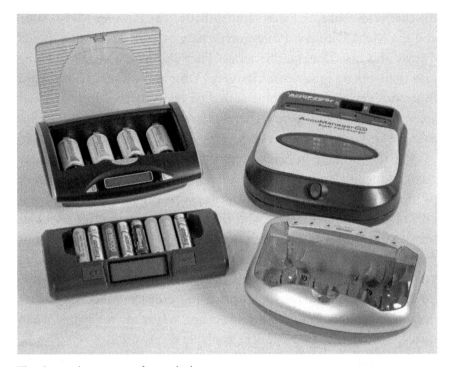

These battery chargers operate from multiple power sources.

reliable flash performance, first responders needing lifesaving portable two-way radio communications, and field technicians needing reliable battery-powered test equipment soon learn which rechargeable batteries give the best service.

Based on their recommendations and my own experience, I have found the Panasonic and Sanyo "Eneloop" rechargeable batteries consistently rated highest for charge capacity and reliability. There may be other brands equal to or better than these, so check mAh ratings before making a purchase. If you are testing batteries with a voltmeter or battery tester, keep in mind that all nominal 1.5-volt Ni-MH batteries will actually measure 1.2 volts, not 1.5 volts, when fully charged.

The AccuManager 20 charger by AccuPower has four charging slots plus two 9-volt battery spaces. The Ansmann Deluxe Energy 8 charger has eight charging slots plus a really nice digital display that indicates the charge level of each battery individually. The Powerex MH-C800S and the Maha Ultimate Professional chargers both have a nice digital display and space for eight batteries. They also have a higher charging rate than other, lower-cost chargers. Your cell phones, iPods, and Bluetooth devices usually come with their own chargers, so you should keep all chargers and spare rechargeable batteries at a central location to make replacing batteries easier.

For the ultimate in being prepared, some models of multibattery chargers include both a 120-volt AC power adapter and a DC adaptor to fit the 12-volt utility outlet in your car or truck. Since the 12-volt DC adapter will easily mate with the power connection on a foldout solar charger, this will allow keeping all of the smaller batteries for your flashlights and small electronic devices charged over and over again using grid power, generator power, your vehicle, or the sun. This is one grid-down preparation you can do now that will provide immediate savings in battery replacement costs while getting you ready for a time when disposable batteries are no longer available at any price.

Powering a home from a rechargeable battery bank is not a new concept. This was reality for hundreds of thousands of homeowners before the electric grid was extended in 1936. However, if this country experiences a true grid-down event, the only lights and electrical devices that will still be working long after the backup generators run out of fuel will be battery powered. I am assuming whatever caused the actual grid-down event was not an electromagnetic pulse (EMP), which could destroy all electronic devices if not stored in an EMP-proof storage container, which will be discussed in chapter 22.

For years I have been advising preppers to own several sizes of fold-up solar chargers and at least one large solar module and deep-cycle battery. Then it finally dawned on me: most people do not own anything that could be recharged with a solar charger even if they had one! Cell phones and the TV remote control are about the only electronic devices in a typical home that are battery powered. While a flashlight should be included in the mix, most non-preppers can't even find a flashlight in an emergency, let alone one that actually works.

Consider the endless variety of battery-powered devices now available thanks to the boating and recreational vehicle (RV) industry. It's time for the prepper community to consider life after generator, and what I am about to propose is far easier to achieve and much less expensive than building a solar home to power conventional 120-volt AC appliances and lighting circuits from a central battery bank, inverter, and a roof full of expensive solar modules.

Not only can a cell phone, flashlight, and portable radio be recharged repeatedly using the sun, but so can many other very useful electronic devices during a grid-down event, and without a generator.

When utility power is months away from being restored, it's much easier to recharge a battery-powered electronic device than to power the same 120-volt AC device using a generator or whole-house solar system. Most 120-volt AC appliances and electronic devices are not designed to minimize their energy usage, since energy efficiency adds to product cost. Saving a few watts of energy is not worth the extra product cost to most buyers when an endless supply of fairly low-cost electricity is available from any nearby wall outlet.

While today's appliances usually display an Energy Star efficiency rating tag, even the most energy-efficient 120-volt AC appliances have an energy usage that is still too high for our grid-down needs. For example, I have tested many satellite receivers and audio-video components that include a remote control. Most measured almost the same electrical wattage load, regardless of being turned on or off with their remote control. Designers of battery-powered devices do not have this energy waste luxury. If their products do not provide an acceptable period of time between charging, the marketplace will quickly toss them aside. One of my first digital cameras was powered by four AA batteries and ate through batteries. When doing photo documentation for site visits, I had to carry a pocketful of spare batteries to make it through the day. These cameras quickly lost out to newer models having new types of rechargeable batteries that allowed taking hundreds of photos before needing a charger.

LIGHTS ON

Some battery-powered devices available

Most of today's battery-powered tools and consumer electronics utilize the latest battery technology, which provides far more runtime than previous battery types, and charging is also much faster. No longer do you need to charge your camera or cell phone all night. Fast chargers having a much higher amp output are available that are capable of recharging these newer batteries in two hours or less. Chargers and charging adaptors are now available that allow recharging battery-powered devices almost anywhere, which is a real advantage when barricaded in your home or bugging out during an emergency. Any battery-powered tool, television, or medical device can be recharged from any car, truck, boat, RV, generator, or a working electric grid, as long as you have the right adaptor and cable.

There are now battery-powered cell phones, pagers, iPads, laptop computers, flat-screen televisions, DVD players, AM/FM radios, GPS receivers, digital cameras, shortwave radios, 2-meter transmitters, walkie-talkies, fans, LED flashlights, LED lanterns, stereos, and electronic games.

INTRODUCTION TO BATTERY POWER

In addition, both 120-volt AC and 12-volt DC chargers are available to power all kinds of battery-powered tools. These include rugged rotary drills, power screwdrivers, hammer drills, circular saws, reciprocating saws, hacksaws, jigsaws, job-site radios, and even the recently introduced battery-powered chain saws. By standardizing on the same brand of tool, they all can use the same interchangeable battery packs, making it easier to have at least one or two fully charged batteries at all times.

The next few chapters will address each type of battery-powered device and technology you will need, with recommendations as to selection and operation. I think you will be surprised as to what is available, and how important each will be for a true prepper.

5

UNDERSTANDING SOLAR SYSTEM COMPONENTS

MOST OF THIS TEXT DESCRIBES HOW solar power should be a part of any long-term grid-down preparations. However, most large-scale solar backup power systems are expensive and will require a qualified dealer to install. I have also pointed out in multiple chapters how a small appliance or electronic device can be powered indefinitely by a small-scale solar system that can be safely installed by anyone having hand tools and a basic understanding of electrical wiring.

Most of the grid-powered electrical appliances and electronic devices in your home can also be powered by a generator during a power outage. However, during a true grid-down event, the power outage most likely will outlast your fuel supply. While it is possible to convert a conventional home to total solar power, most people will not have the budget, equipment space, or roof orientation for this approach.

By looking for the most efficient models when purchasing your

Total Off-Grid Solar Home

appliances and electronic devices in the first place, you can substantially downsize any backup power system and save far more in operating cost than the additional price of the more efficient models.

Many everyday appliances and electronic devices are available in a smaller size with lower energy requirements that can operate directly from a 12-volt battery when both grid power and generator power are not available.

The following sections will describe the function of each major electrical component needed to install a do-it-yourself solar backup system, and how to determine specifically what components you will need. Keep in mind any poorly installed electrical system can burn down your home and even kill, so seeking the assistance of a licensed electrician is strongly recommended.

A. SELECTING A SOLAR MODULE

A solar module is a manufactured assembly, usually having an aluminum frame and a tempered glass or flexible clear plastic glazing, covering a fixed number of individual solar cells and their interconnect wiring, which is enclosed on the back with a vacuum-sealed vinyl sheet. The back includes a junction box to make electrical connections to the internal positive (+) and negative (−) terminals.

Most higher-wattage modules intended for large-scale multimodule installations have sealed junction boxes and two molded interconnect cables having polarized male and female connectors. Solar cells generate an average of one half volt at full sun exposure, so the number of individual cells in any solar module that are wired in series determines the maximum voltage of each specific module.

500-watt solar array using homemade supports

For example, most 125-watt and smaller solar modules have an open circuit voltage (not connected to any load) of 21 volts DC. When connected to an external load, these modules will operate around 17 volts, which is ideal to charge a 12-volt battery.

Most solar modules over 125 watts will normally have an open-circuit voltage of 35 volts DC, and will operate around 28 volts when connected to an external load, which could be used in 24-volt battery backup systems. However, most higher-voltage modules are normally wired in series with other modules to produce over 300 volts DC to supply an inverter selling power back to the grid and not charging batteries.

The biggest mistake most novices make is trying to use these higher-voltage solar modules designed for grid-tie systems in a 12-volt DC battery system. This will quickly burn up any 12-volt electrical components and destroy the battery by overcharging and drying out the electrolyte.

The three major types of solar modules available to the general public are *single-crystal, polycrystal,* and *amorphous.* Single-crystal and polycrystal modules have a slightly different color and appearance but are almost electrically identical, and either can be used for the solar projects described in this book. The single-crystal module will have solid dark-gray-colored solar cells, while the polycrystal will look like galvanized metal, except blue in color. However, the amorphous solar module is less than half the efficiency of the other two, so while it is by far the lowest-cost unit, it will take two to three times the roof area to generate the same charging power as a single-crystal or polycrystal module array.

The amorphous solar module is easy to spot since the process involves "plating" the silicon onto the back of the glass in a continuous layer, not separate, individual solar cells. They are dark gray or black in color and are typically used in handheld solar-powered calculators and solar walkway lights, since they are very inexpensive to manufacture.

All solar modules are required to have a label identifying the brand, model number, country of manufacture, and performance under standard test conditions (STD). This is basically the solar radiation found at sea level, in a southern desert, at high noon, with a clear blue sky, with the module facing directly into the sun, and with low-ambient air temperature.

While this forces all manufacturers to be on the same level playing field using the same test conditions, I assure you it will be very rare that your solar system will ever achieve more than 80 percent of the wattage indicated on this nameplate. This is especially true for any solar module

SOLAR MODULE LABEL EXPLANATIONS

TERM	INDICATES	NOTES
Voc	OPEN-CIRCUIT VOLTAGE	VOLTAGE WHEN NOT WIRED TO LOAD
Vmp	MAXIMUM POWER VOLTAGE	VOLTAGE AT MAXIMUM WATT OUTPUT
Isc	SHORT CIRCUIT CURRENT	MAXIMUM AMP FLOW AT SHORT CIRCUIT
Imp	MAXIMUM POWER CURRENT	AMP FLOW AT MAXIMUM WATT OUTPUT
Pmax	MAXIMUM POWER	MAXIMUM WATTS AT TEST CONDITIONS
Fuse	MAXIMUM SERIES FUSE	MAXIMUM AMP FUSE ALLOWED
Vmax	MAXIMUM VOLTAGE	MAXIMUM VOLTAGE ALLOWED IN ARRAY
STD	STANDARD TEST CONDITIONS	NOTE LISTING TEST CONDITIONS USED

having a fixed mounting or less than perfect orientation, which may drop this rating even lower.

The label will also indicate several voltages and several amp currents. The open-circuit voltage (Voc) indicates the voltage at the terminals when in full sun and nothing is wired to the output terminals. This is helpful to make sure all wiring insulation and downstream charge controllers are rated for this maximum voltage. The maximum power voltage (Vmp) is the voltage the module needs to produce the highest wattage possible.

The short circuit current (Isc) is the amps generated at the module terminals in a dead short. This is the value used to size the wire between the solar array and the charge controller or grid-tie inverter and is multiplied by a 1.56 percent safety factor. (See appendix table 3 for example.) Finally, the maximum power current (Imp) is the amp flow the module needs to produce to reach the indicated wattage. Some solar charge controllers will vary one or both values to force the module to operate at this maximum power point.

You want to make sure the product label includes a recognized safety trademark, which will typically be Underwriters Laboratories (UL) or

Canadian Standards Association (CSA). If any of this information is missing, you may be viewing a poorly constructed module or a module made with inferior materials that may work for a few years but then fail after exposure to repeated hot summers and cold winters or humidity. I have actually seen a row of solar modules catch fire and totally melt down into a puddle of molten glass and aluminum, then partially melt through the roof surface. Sometimes extra-low or close-out solar module prices are low for a reason.

In preparing for a grid-down event, I strongly recommend purchasing only single or polycrystal solar modules, and make sure they have a junction box on the back that can be opened. Higher-voltage modules having molded interconnect cables or the low-efficiency amorphous modules will require too many modifications of other system components and will require cables with expensive weatherproof plugs to use in grid-down battery charging systems and should be avoided. For remote off-grid solar arrays under 1200 watts, a pole-mounting array takes up minimum ground area and keeps the module wiring away from animals.

Wind loading on pole-mounted arrays larger than 1200 watts can be excessive. To avoid needing a much larger support pipe and concrete foundation, it is usually more practical on higher-wattage applications to install two or more separate pole-mounted arrays of 1200 watts each, or switch to a roof or ground array mounting.

I have found a ground-mounted array located near the home or cabin is much easier to install, and avoids the potential of roof leaks and future roof replacement

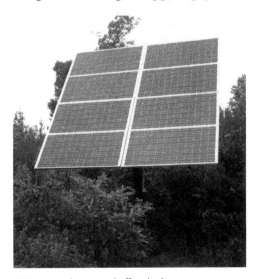

1200-watt pole-mounted off-grid solar array

problems if you can protect the back wiring from animal damage and kids using it for a jungle gym.

B. SELECTING A SOLAR CHARGE CONTROLLER

A solar charge controller is similar to the voltage regulator in a car that takes the constantly changing voltage and current output from the alternator and controls the charging power going into the battery. All solar-power systems with a battery will require a solar charge controller. The lowest-cost charge controller will not have any meter display, and works basically like an on-off switch. It has a positive and negative connection for the solar module, and a positive and negative connection for the battery. Its only function is to automatically allow solar electricity to

Basic and PWM-type solar charge controllers

pass through to the battery when the battery voltage is lower than the solar array voltage, and turn off again when the battery voltage is high.

It also will block any battery charge from leaking back through the solar modules at night. Although low in cost, this basic charge controller cannot adjust the voltage or current flow going into the battery charging. The charging voltage and current are totally dependent on the balance of voltage and current between the solar module and the battery, and this will vary during the charging process based on time of day, ambient

UNDERSTANDING SOLAR SYSTEM COMPONENTS

temperature, cloud cover, and battery voltage.

The slightly more expensive pulse-width modulation (PWM) charge controller is also basically an on-off switch wired between the output terminals of the solar module and the battery. However, this charge controller monitors the battery voltage. The full charging voltage from the solar array reaching the battery is constantly being turned on and off (pulsed) at a different rate and length (time) of pulse based on the charge level of the battery. By slowing down the charging process as the battery reaches its full charge state, this PWM charge controller prevents overcharging and overheating the battery as it reaches a full charge state.

The maximum-power-point tracking (MPPT) charge controller is the most-efficient solar charge controller currently available and includes multiple program features. Since this charge controller has control functions that will alter the charging process based on battery type, battery temperature, and battery state of charge, it will provide the fastest charging with the least risk of battery damage from overcharge or overheating.

In addition, an MPPT charge controller separately adjusts the voltage and current leaving the solar

Programmable MPPT-type solar charge controllers

module, which forces the solar array to operate at its point of maximum output wattage, while separately controlling the charging rate the battery requires to safely reach a full charge in the shortest time possible.

This means an MPPT charge controller can operate with a much higher solar array voltage than the battery voltage. This allows multiple solar modules to be wired in series with combined voltages in the 80- to

150-volt DC range, to charge a 12-, 24-, or 48-volt DC battery bank. While most off-grid remote cabins may not need more than 12 volts DC to power small lights and DC appliances, being able to wire the solar array for much higher voltages means it is possible to locate the solar array farther away from the house or cabin without the need for large wires since the voltage drop between the solar array and the home will be much less as the array voltage is increased.

An MPPT charge controller will allow using higher-voltage and higher-wattage solar modules to charge a lower-voltage battery, since any difference in voltage is converted into more charging amps by this controller and not wasted. Selecting a charge controller can be confusing, but you usually get what you pay for, so for larger solar arrays and battery systems, a more expensive MPPT charge controller will force the system to collect and store a much higher percentage of the available solar energy. On small solar battery charging systems, a lower-cost PWM-type charge controller is usually adequate.

C. SELECTING A SOLAR BATTERY

While battery technology has undergone some major advances to increase power density and reduce weight, thanks to electric car research, battery weight and physical size are usually not as important in a residential off-grid solar and emergency backup power system.

For most small residential solar systems, the most common battery with the least cost is the 6-volt T-105 (golf cart) battery, and the larger-capacity L16 industrial battery, which is the same length and width as a golf cart battery, but almost twice as tall. The 6-volt golf cart battery weighs an average of sixty-three pounds and can store approximately 1 kWh (kilowatt-hour) of energy, while the L16 battery weighs more than one hundred pounds and can store 2 kWh of energy. Both have thick plates and are designed to handle the large daily swing in charge level typical with an off-grid home. Of course, using 6-volt batteries will require multiples of two for a 12-volt DC power system.

The major disadvantage with these "wet cell" batteries is having to add distilled water every three to six months, depending on their age and depth of daily discharge. Any battery requiring more distilled water every few weeks is a sign of either overcharging or that the battery has reached its end of life. They also will produce hydrogen gas as they near a full-charge state if the charger does not back off the charging rate, and this gas can be explosive if allowed to become concentrated in a confined area.

Any battery bank using deep-cycle wet cell batteries should be located in a well-ventilated area and away from open flames. Locating

Sealed AGM, gel, and liquid deep-cycle solar batteries

these in a garage or storage shed having a concrete floor is an advantage due to the heavy battery weight and potential for spilled acid during overcharging and refilling. However, keep in mind the ideal battery temperature is 77 degrees Fahrenheit, so in cold climates the battery amp-hour storage capacity can be substantially reduced as the battery plate temperature drops below 30 degrees Fahrenheit.

Off-grid battery bank using L16 batteries

It's recommended to support deep-cycle batteries on several raised pressure-treated boards resting on concrete blocks instead of directly on the floor, since concrete floors in unheated garages stay much colder than the ambient air temperature, which would reduce the amp-hour capacity of the batteries.

For mobile applications and when a solar-charged battery must be located inside the home or cabin, a sealed AGM or gel battery is your best choice. These are much more expensive than a wet cell battery without providing any additional life or charge capacity. However, a sealed battery does not normally produce explosive gases during the charging process, as long as the charger is designed to charge a sealed battery without overheating or overcharging.

Sealed batteries require a slightly lower charging voltage than when charging a wet cell battery. In addition, a gel/AGM battery contains no liquids to leak and can be mounted or carried upright, sideways, or even upside down. Generally, the gel battery is used for applications having a more robust daily charge and discharge cycling, while the AGM battery is typically used for applications having a longer period of no load between an occasional heavy load followed by a slow recharge.

By far, most of the gel and AGM batteries are sold in 12-volt versions in group sizes 24, 27, and 31 for RV and boating applications. They are also available in the 6-volt group size GC-2, commonly referred to as the T-105 golf cart battery, which can handle heavier deep cycling than a 12-volt battery, but are more expensive. For really large battery-power needs and when a sealed battery is not required, many totally off-grid solar retreats use the larger 6-volt group L16 battery, with eight or sixteen batteries making up a typical off-grid battery bank.

> The Battery Council International (BCI) is the main trade association for battery manufacturers, and the BCI group numbers identify the type of battery. Here are a few main groups and their amperage per hour rating.

BCI GROUP	L"	W"	H"	AMP HOURS
6-VOLT				
GC2	10.38	7.13	10.88	225
L16	11.64	6.95	15.73	400
12-VOLT				
22NF	9.4	5.5	8.9	55
24	10.3	6.8	8.9	80
27	12.1	6.8	8.9	90
30H	13	6.8	9.5	110
31	12.9	6.75	9.27	125
4D	20.5	8.13	10	200
8D	20.75	11	11	245
34	10.3	6.8	7.9	800

All batteries are sized based on amp-hours of capacity, which will be different depending on the hours the load is being powered. For off-grid power systems, use the twenty-hour rate table data. When comparing brands, be sure to use the same hour rating data. For example, a battery with a 100-amp-hour rating may power a 10-amp load for ten hours, but the twenty-hour table may show a 110-amp-hour capacity when the discharge rate is slowed down. Also compare battery weight, as this will vary from manufacturer to manufacturer for a given battery group size. Usually the more weight, the more lead was used in the cell plates, which can increase amp-hour capacity and expected life.

Batteries should not be discharged over 50 percent, as each total discharge will significantly reduce battery life. Good design practice will dictate cutting all advertised amp-hour ratings in half to come up with the maximum daily discharge and recharge you should apply to any battery. Refer to the battery dimension and wiring configuration charts in the appendix for more detailed information.

D. SELECTING AN INVERTER

The inverter converts a 12-volt DC battery voltage into 120-volt 60-cycle AC to power any lights and electronic devices you cannot find in 12-volt DC versions. While these come in all sizes and with endless options, for smaller off-grid and grid-down applications you will usually need only an inverter with a 300-watt rating to power a flat-screen television, laptop computer, satellite receiver, or cell phone. Inverters this small are normally designed to plug into a vehicle's 12-volt DC utility outlet, which is perfect for most mobile and bugout power requirements.

Larger inverters will require a direct power connection to the vehicle's battery, and this wire can be fairly large if the inverter wattage exceeds 1200 watts. Above 2400 watts most inverters are designed to be powered by a 24- or 48-volt battery since a 2400-watt load at 120 volt AC translates to a 200-amp load on a 12-volt battery!

Any electrical heating appliance and any major household appliance, including refrigerators, washing machines, and air-handling units,

Grid-tie and off-grid residential inverters

UNDERSTANDING SOLAR SYSTEM COMPONENTS

would quickly discharge a single 12-volt battery. These large loads should be used only when generator power is available during a grid-down event and not powered from an inverter and battery.

In addition to different wattage ratings, inverters are advertised as having either a "modified" or "pure" sine wave output. The pure sine wave inverter produces a 120-volt 60-cycle AC output that is almost exactly the same power quality as grid power. Any television, electronic device, or battery charger will function normally on a sine wave inverter.

A "modified" sine wave inverter requires fewer parts to manufacture and is much less expensive than a pure sine wave inverter. However, the 120-volt AC output of a modified sine wave inverter is actually a series of voltage "steps," which approximates the 60-cycle sine wave profile of the utility grid, but not as smoothly. While most lighting fixtures and motor-driven appliances will work normally on a modified sine wave inverter, more sensitive electronic devices may have problems.

For example, the television display may not fill the entire screen, some microwave ovens may take longer to cook the same food item, and some battery chargers may take longer to recharge a battery when powered by a modified sine wave inverter. Fortunately, since many of today's electronic devices are designed to work with both 120-volt

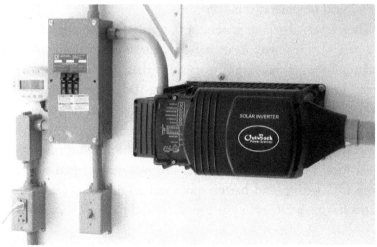

Outback pure sine wave 3600-watt off-grid inverter

United States and 240-volt European electrical systems, their power supplies are much more forgiving of power quality and voltage, and can usually operate just as well on either inverter type.

While a pure sine wave inverter will typically cost two to three times as much as a modified sine wave inverter having the same wattage, there are still some sensitive electronic devices that will work normally only on the pure sine wave inverter, and in many cases you will only know for sure by trial and error.

Inverters designed for mobile application do not normally make good residential off-grid inverters since they do not include a built-in battery charger or automatic transfer switch. Inverters that are designed for a residential application usually have a higher-wattage capacity since they are not limited to a 12-volt vehicle battery voltage. In addition, residential inverters usually have a built-in battery charger and automatic transfer switch.

During normal grid operation or when power is available from a generator, the 120-volt AC power source passes straight through the transfer switch in the inverter and is routed to the loads connected to the inverter's output. The inverter immediately takes over and supplies these connected loads from the battery-powered inverter circuit if external grid or generator power is lost.

When external grid or generator power is available, the inverter will route part of this power to its battery charging function, which has a substantially larger charging amp capacity than a typical separate battery charger. For most 12-volt DC inverters, a 100-amp charging output is typical and can recharge batteries much faster than most automotive-type battery chargers.

Please note most 120-volt AC circuit breakers and fuses typically sold in local building supply stores are not designed to protect low-voltage DC wiring and battery-powered equipment. Use only components having a DC rating in any solar and battery-powered wiring.

E. WIRE AND FUSES

Electricity, especially high-voltage electricity, can kill, and high current can overheat wiring components and cause fires. Keeping this in mind, this section is not intended to cover everything you need to know to wire any electric circuit safely. I always encourage consulting with a licensed electrician if in doubt.

Wiring low-voltage DC electrical components is far different from 120-volt AC household wiring, and the fuses and circuit breakers sold in your local builder supply for house wiring are not safe to protect DC electrical circuits.

When wiring any electrical component, you first need to determine its maximum amp load, which is usually indicated on the nameplate and in the owner's manual. For example, the owner's manual for an inverter may indicate the maximum load on the batteries will be 93 amps, but recommends using a 100-amp fuse or circuit breaker.

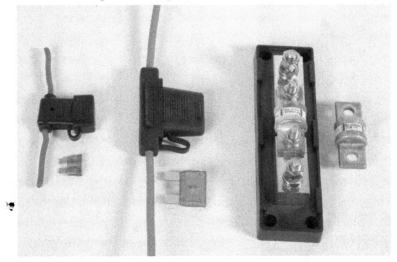

DC-rated in-line 30-, 80-, and 150-amp fuse holders and fuses

The next step is selecting the size of wire needed to connect this inverter to a battery. Although the *National Electric Code* has detailed wire sizing tables that include adjustments for temperature, number of conductions in same conduit, and wire insulation material, in general a No. 10 wire is normally used for a 30-amp circuit, a No. 6 wire for a 60-amp circuit, a No. 2 wire for a 100-amp circuit, and a No. 2/0 wire for a 150-amp circuit. In addition, as the distance between the load (inverter) and the power source (battery) increases, the wire resistance will cause a significant voltage drop and you may need to go to the next larger wire size. However, you do not need to increase the fuse size, as this wire size increase is to reduce voltage drop, not increase the amp load.

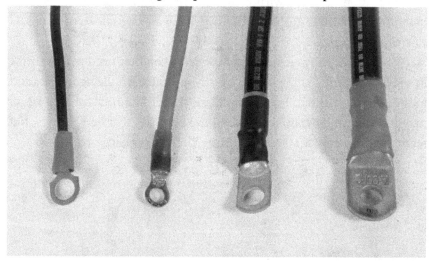

Typical No. 10, No. 6, No. 2, and No. 2/0 wire and battery terminals

For example, if you were installing wire from your house to power a new 120-volt AC floodlight one hundred feet away, the actual voltage reaching this light fixture might be only 116 volts due to the wire resistance. This 4-volt drop, which is a 3 percent voltage loss, would most likely have little or no effect on the operation of the light, which will probably be rated for 110- to 120-volt operation.

However, if this were a 12-volt DC floodlight having the same wattage bulb, this 4-volt resistance drop, which is a 34 percent voltage loss in the one hundred feet of wire, would mean only 8 volts makes it to the light, which is below the voltage limit for this fixture. Keep all DC wiring as short as possible, and use only copper wire having a heavy insulation. Large loads, including inverters, should be located no more than ten feet away from the battery, with five feet even better. Make sure any wire connected to the positive (+) terminal of the battery includes a fuse, which should be located as close as possible to the battery post for maximum wire protection.

The best types of deep-cycle batteries for off-grid and solar power systems will have bolt-type "flag" terminals, not automotive-type round battery posts. The preferred method to connect larger-size battery cables to bolt-type terminals will be heavy copper lugs, crimped on the end of flexible copper cable. While heavy welding cable is flexible copper and designed to carry high amp flows, welding cable is not approved for house wiring due to the smoke rating of the cable's insulation.

At high-current flow, any poor electrical connection will overheat due to the higher resistance. Just before making final connection to any battery post, be sure to wire brush both the battery post and the wire's terminal to a bright and clean condition to minimize resistance, as lead oxidizes very quickly. The lead battery post will be bright silver in color when cleaned properly.

A final note concerning the location of DC wiring: Unlike AC wire runs, placing two long parallel conductors down a ditch line or across a ceiling that are carrying DC electricity produces the same effect as a long, narrow capacitor. A capacitor is constructed from two very long strips of metal foil, separated by an insulator sheet, rolled up into a tight cylinder. Electrically, these can store power briefly and also build up an electrical charge that is in reverse to the normal electric flow.

Each of the two parallel wires carrying DC electricity has its own magnetic field surrounding each wire whenever DC electricity is flowing, and these magnetic fields can cause additional resistance in the wires due to this "capacitance effect." This can generate significant radio interference in any nearby radios or shortwave radio equipment.

However, it is very easy to force each of these separate fields to cancel each other out and minimize any radio interference. Just make sure the two insulated wires are touching along their full length, which can be done with wire ties to eliminate any gap. While reducing the gap between the wires helps, you can almost eliminate the problem completely if you twist the two wires together along their length as they are being installed. This twisting together is not as easy as it sounds for very long or very large wire sizes, but the number of turns per foot does not need to be excessive, and you do not want to stress the wire insulation.

When burying any underground wire, I always partially backfill the ditch to cover the power wiring, and then lay a strip of bright plastic survey tape down the ditch line near the surface before backfilling the remaining depth. This is very cheap insurance, as all backhoe operators know to stop digging when their bucket pulls up a long yellow or pink plastic "streamer" hanging from the bucket! If you are like me, you may forget where all of the underground utilities are located around your home, and "Miss Utility" will not mark underground solar wiring between a solar array and the house or barn since it is not considered a utility.

The *National Electric Code* specifies how deep electrical wires must be buried and this depends on several design factors. However, for low-voltage solar wiring, twenty inches deep is a reasonable goal, with sixteen inches the minimum to avoid nicking these underground wires when planting a garden or sinking a fence post. You may be able to go shallower when the ground is very hard by burying the wires in nonmetallic electrical conduit. This conduit is very inexpensive and easy to install.

Any DC solar wire installed below ground that is not in conduit must be underground service entrance (USE) rated. Table 3 in the appendix provides ratings for the most common wires used to install solar equipment. Typically, the guidelines for underground solar wiring are usually a concern only when installing a ground or pole-mounted solar array, not a roof-mounted array.

6

LIGHTING WITH BATTERY POWER

NOT COUNTING FIRE FOR SPACE HEATING and cooking food, having an adequate source of light is a primary power requirement, whether you are at home, camping, hiking, or bugging out. Without artificial lighting, your lifestyle will be limited to only those tasks you can do during daylight hours, and in rooms having windows. In addition, without artificial lighting during a dark night, the areas around your home, campground, or trail will be extremely dark and unsafe.

You will need some light in every room that you occupy on a regular basis during a power outage, but may not need the higher light levels you require during normal times. Areas for meal preparation usually require the highest light levels. With the exception of using lower-wattage four-foot T8 fluorescent ceiling fixtures in these areas, switching other room light fixtures to lower-wattage LED bulbs and using table lamps for task lighting will significantly extend the number of days you can operate from any backup power system. You can find an LED bulb for almost every size and style light fixture imaginable, including 12-volt DC-powered fixtures for RVs and truck campers if you do have to bug out.

Superefficient LED and CFL lights

To prepare for a true grid-down event, replace all incandescent lightbulbs located in high-occupancy areas with LED or compact fluorescent lamps. Although more expensive, replacing all 60-watt or larger incandescent lightbulbs with LED lamps means your backup power system just got downsized by a factor of ten, or will provide the same light levels for ten times longer. This will also reduce your current monthly electric bill while utility power is still available.

LIGHT FIXTURES

When selecting light fixtures for use during a grid-down event or to minimize electrical usage when still on the grid, avoid light fixtures having a frosted lens and nonreflective interior. Look for fixtures having a clear glass lens or bulb enclosure and polished metal or mirror-type interior to maximize the light being projected from the fixture. This will allow using lower-wattage bulbs and is critical when using LED-type lamps.

SOLAR PATHWAY LIGHTS

Solar-powered LED pathway lights can be found in almost every yard and driveway these days, yet most people don't realize how these make perfect emergency lighting systems during a power outage. They are charging outside in the sun all day, and then they provide light throughout the night. Marketers advertise X as a way to rate LED walkway lights lumen output. One X equals one lumen of light, which is equivalent to the light from a single candle as measured at a foot away.

You will need to shop for the larger solar walkway lights having a 24X or higher rating for the kitchen, dining room, and family room. Smaller solar walkway lights in the 6X capacity range should provide adequate light levels in halls and stairwells. Bathrooms will require 12X or higher for adequate lights at sinks and mirrors. Make sure the models you select have clear lenses and polished-mirror reflectors to

Solar pathway lights make great emergency indoor light fixtures.

maximize the light distribution. Since you will be buying these for the additional purpose of providing emergency backup lighting, avoid models with a lower capacity battery and flimsy construction.

Although designed to be installed in the ground with a spiked tube support, a more practical installation involves driving a three-fourths-inch PVC conduit in the ground with two feet remaining exposed. Be sure each location provides the normal exterior illumination you need, while being free of shading during the day. Of course you should make sure the outdoor location selected is also near the house to help reduce risk of theft. Remove and discard the spike end and insert the tube support of each solar light into the conduit. This makes it easy to remove and take inside each evening during a power outage.

Using discarded wood blocks, mount a short piece of the same size PVC conduit into the center of each wood block, creating a lamp base. You will need one for each room requiring emergency lighting. Each morning, simply remove the solar lights from their wood bases and return to the yard for recharging. This is an ideal power outage task for the kids. Even if you still have fuel for your generator, there is no need to waste it when this simple do-it-yourself project can provide all the interior lighting you need during any power outage, regardless of how long it lasts.

The rechargeable batteries that come with the more expensive solar pathway lights typically have a two-year life under normal usage. Try to standardize on models requiring the same size batteries, and be sure you have spares.

FLASHLIGHTS

Every home should also have several LED flashlights that are fully charged and located for easy access in the dark. LED flashlights that are powered by rechargeable batteries are a must-have. Chapter 4 addressed the benefits of today's longer-lasting rechargeable batteries. During extended power outages, generators and emergency backup systems may fail or need to be shut down for service or to save fuel. When this happens the house will get extremely dark, and this can occur unexpectedly. Having an LED flashlight beside every bed and main entry door is critical to your grid-down planning. You will also want an LED flashlight in each vehicle and each bugout bag (see chapter 17). A word of warning, however: I have found that most of the lower-cost LED flashlights require three of the smaller AAA batteries, while the slightly larger and more expensive LED flashlights will use two AA batteries.

Throughout this book I stress the importance of using rechargeable batteries and making sure all of the battery-powered devices you own require a limited number of battery sizes. You should avoid any device that requires the small, AAA batteries, as these are just too small to power anything for any length of time. Probably the most useful battery size that fits almost all digital cameras, flashlights, electronic games, and portable radios is the slightly larger AA battery.

LED LANTERNS

Originally sold for the camping and RV enthusiast, my favorite room light during an extended power outage is an LED lantern similar in design to the old propane or kerosene lanterns, but smaller. Unlike a single-direction flashlight, the LED lantern is designed to project its light in all directions, and most models have a much larger battery capacity than a typical flashlight. Some models include a fold-up hook for hanging, and can be easily carried from room to room and hung from a ceiling hook.

When suspended near the location of any existing ceiling light fixture, a single LED lantern will provide acceptable lighting levels throughout most residential-size bedrooms, family rooms, and bathrooms. Adding a small hook near each ceiling fixture to hang these lanterns during a power outage is an easy way to prepare now, and people most likely will never notice the hooks.

When selecting an LED lantern, make sure it uses the same size rechargeable batteries as your other battery-powered devices (C batteries are the most common batteries used in the larger LED lanterns). If the

Solar-charged and battery-powered LED lanterns

batteries are designed to stay in the lantern and be recharged from a 120-volt AC power supply, you will need to order the optional 12-volt DC adaptor so they can be charged from multiple 12-volt DC power sources during a grid-down event. Try to purchase the same model of LED lantern to use in multiple rooms in addition to using the same size batteries. At a minimum you will probably want a hook-suspended LED lantern in the family room, kitchen, and each bedroom. Smaller models or solar walkway lights can be used in corridors, bathrooms, and stairwells.

During any extended power outage, having some light at night in each room can make a difficult time at least bearable. Do not think owning one or two flashlights will do the job either. Your first goal in being prepared after taking care of food and water should be multiple battery-powered LED lights and a way to keep these charged.

7

EMERGENCY COMMUNICATION WITH BATTERY POWER

KNOWING WHAT IS HAPPENING IN YOUR community, state, nation, and world, not to mention staying in touch with family members, is essential during any type of crisis or disaster, and loss of grid power is usually a part of most emergency events. During an extended power outage your grid-powered radios, televisions, computers, Internet routers, satellite systems, and phones will not operate. While today's cell phones offer texting, emailing, and Internet access as well as wireless phone communications, most generators located at each cellular tower will have only a week of emergency backup fuel storage. In addition, most likely your cell phone's battery charge will last only a day, and without the grid your regular phone charger will not work.

Having reliable communications and access to news is not only helpful during a power outage, but it can save your life when you are able to hear extreme weather alerts, travel advisories, health warnings, and civil unrest news. This chapter will review the many types of emergency communications that are available, and multiple ways to use these systems during an extended power outage.

AM/FM RADIOS

The "medium-wave" frequencies of 540 to 1610 kHz are assigned for the AM broadcast radio band in the United States, and the "higher-frequency" band of 87 to 108 MHz is reserved for FM broadcasts. All AM/FM radios sold for use in the United States will at a minimum include these radio frequency bands.

Having a portable battery-powered all-band radio can be a real lifesaver during an extended power outage, and it does not require running a generator.

If you decide to purchase a handheld radio, select a model with both a speaker and ear buds. While the speaker allows others to listen, switching to the lower-power ear buds when alone provides much longer battery life.

While we normally think of a battery-powered radio as something you can carry in the palm of your hand, several tabletop models are available with high-capacity rechargeable batteries and large speakers for room-filling sound. Like the families of the 1920s, who gathered around their battery-powered Delco radios, your family may one day gather around your battery-powered tabletop radio. During a grid-down event, this may be your main source of both news and entertainment when all other forms of electronic communications are down.

Look for a portable radio model that requires either the AA or the larger, C-size batteries. Avoid any radio requiring the tiny AAA or large D flashlight batteries, as it is doubtful any of your other battery-powered devices will use these. Always remember that during a real grid-down event, keeping your battery-powered devices operating and fully charged will most likely require playing "musical chairs" with batteries, adaptor cables, and various charging devices, so keep it simple, and standardize wherever possible.

Since most people today download their music from the Internet or transfer their CDs to their portable devices, the AM/FM receivers built into most home stereo sound systems are probably not being used. However, today's AM radio band is very popular for talk radio shows since an AM radio transmission can extend for hundreds of miles. While FM transmission provides clearer stereo broadcasts that are virtually free of static, FM transmission is limited to line-of-sight distances. Depending on the height of the FM station's antenna, the reception area rarely exceeds twenty or thirty miles, especially in mountainous terrain.

With the popularity of talk radio, manufacturers of battery-powered portable AM radios now include higher-quality AM antennas and more sensitive AM receiver circuits to maximize reception of distant AM stations. In addition, some battery-powered radios will also receive several shortwave radio bands, allowing the reception of overseas programs and news broadcasts typically available in English as well as their country's language.

SHORTWAVE RADIOS

While very few AM/FM radio models will include any shortwave bands, almost all shortwave radios will include the standard AM and FM radio bands, along with many more broadcast frequencies and bands. To make it easier to find a specific station, the large frequency spectrum is divided into specific "bands" typically referenced in meters. In reality this is the physical measurement of the "length" of the radio wave from start to end of each cycle. By referencing the sidebar showing frequencies versus wavelength, this means a station broadcasting at 1.820 Mhz will have a wave that will be found on the 160-meter radio band, be 165 meters (541 feet) in length, and will complete 1,820,000 cycles every second. Under normal atmospheric conditions, the size of the wave determines how far away you can hear a radio station. A longer wavelength will easily be reflected by the upper ionosphere and "bounce" back to earth many hundreds of miles from the transmitter antenna, while very high frequency waves having a meter length of only a few feet will travel in a straight line, go straight through, and not be reflected by this upper ionosphere. For most frequencies in the two- to ten-meter band, the broadcast distance may be only a few miles away, depending on the height of the antenna.

POPULAR SHORTWAVE RADIO BANDS

METER	COVERS FREQUENCIES
160	1.80 TO 2.00 MHZ
80	3.80 TO 4.00 MHZ
40	7.175 TO 7.300 MHZ
30	10.10 TO 10.15 MHZ
20	14.225 TO 14.350 MHZ
15	21.275 TO 21.450 MHZ
12	24.93 TO 24.99 MHZ
10	28.30 TO 29.70 MHZ
6	50.10 TO 54.00 MHZ
2	144.10 TO 148.00 MHZ

Some shortwave radios may also include several very high frequency bands with wavelengths measured in centimeters, not meters, which definitely is a line-of-sight transmission and requires a very tall antenna to reach over a few miles. These very high frequency bands may include aircraft, marine, police, and fire and rescue channels, although many cities use encryption, making it difficult to receive.

CELL PHONES
Cell phone companies have increased their on-site generator fuel storage based on recent experience with longer storm-caused power outages, but

sooner or later they will run out of fuel, and fuel delivery will be problematic during an extended power outage. Keeping your cell phone fully charged is extremely important, as most cell phones now include the ability to send and receive text messages, which usually can get through during limited reception conditions when normal voice calls will not.

Battery-powered shortwave radios

A quality cell phone may offer other helpful forms of communication during an emergency, including accessing the Internet for web-based alternative news sources, sending and receiving e-mail to distant relatives, and photo documentation of events.

While all cell phones come with a standard 120-volt AC charger, it's critical that you purchase several optional 12-volt DC car chargers and place one in each vehicle and in your bugout bag. While the cellular service will eventually end during a grid-down event, it should still be available in most regions of the United States for the first few weeks unless the event was triggered by an EMP attack—then all bets are off.

Cell phone adapter to fit foldout solar module

WALKIE-TALKIES

Once all cell phone and dial-up phone service has been lost, you still need to be able to communicate with others. The Family Radio Service (FRS) walkie-talkies utilize wavelength channels that are in the ultra-high frequency range of 462 to 467 MHz, and utilize low-noise FM transmission, which has less interference than AM radios. An FRS radio does not have the range to let you talk to somebody in the next city. However, these are a great way to set up a local neighborhood watch during periods of civil unrest, and they are very inexpensive.

Although first set up in 1996 for private family and farm use before the wide use of cell phones, businesses found these handheld radios to be a perfect way to communicate with employees and service technicians working in large warehouses and on multibuilding campuses. These walkie-talkies have multiple channels and very clear reception, and they operate similarly to a cell phone when using their privacy paging tones.

FRS and GMRS walkie-talkies with car charger

While limited in range, a quality FRS walkie-talkie will at least let you talk to nearby neighbors as there is safety in numbers, especially when there is a risk of roving gangs. Since the FRS walkie-talkies are limited to a half watt of transmission power, they do not require a license. While FRS radio advertising typically indicates a range in miles, this is true only over a flat open field or over water. This low wattage limits their range to just one or two city blocks unless you are in extremely open terrain.

The lower-power FRS radios are typically sold in sets of four or six, which avoids the problem of your neighbors using their own radios that may not have the same channels or paging codes as you are using. Being able to hand out battery-powered walkie-talkies with matching preprogrammed channels to neighbors during a grid-down event creates a great local early-warning system.

The demand for longer-range commercial communication has resulted in a more powerful line of extremely rugged handheld radios. The General Mobile Radio Service (GMRS) walkie-talkies operate on the same first seven channels of the fourteen-channel FRS radio band, plus an additional eight channels immediately above the FRS channels. These more expensive GMRS radios have up to five watts of transmission power and do require a station license. However, this license does not require passing a test since it's only a transmitter registration process. These GMRS walkie-talkies provide several miles of very clear communications, depending on antenna selection and whether communicating with a base station or another handheld unit.

CB RADIOS

Citizens band (CB) radio is another excellent form of short-range communication during extended power outages and when cell phone and other forms of communication are down. Initially popular for truck-to-truck communications during the 1970s and before cellular service, the CB radio suffered from too many people trying to talk on a limited number of channels. However, during a grid-down event it can be an excellent way of communicating with others located beyond the range of FRS and GMRS walkie-talkies.

Since CB radios are installed in almost every tractor-trailer rig, they are designed for 12-volt DC power and are extremely rugged to withstand the vibration. Since older models were manufactured before the introduction of microelectronics, these are considered to be less susceptible to EMP damage than newer models having integrated circuit devices. While this may or may not be true, keeping a CB radio in a metal EMP shielding container until needed in an emergency can help guarantee survivability.

Operating in the original CB radio band of 27 MHz (eleven meters) required a station license, an FCC-issued call sign, a 5-watt limit on transmitter output, and a sixty-foot limit on antenna height. Since few truck drivers followed these early call sign and station license requirements, these rules were eventually eliminated. However, to help identify each person on the air these days, everyone is encouraged to use the letter *K* followed by their initials and local zip code to make up their own call sign. The FCC later expanded the original twenty-three channels to forty channels, but lowered all new CB radios to a 4-watt maximum transmit power in AM mode, or 12 watts in single-sideband (SSB) mode.

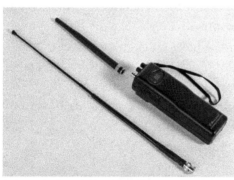

A 4-watt 40-channel CB walkie-talkie

Since CB radio is still a very popular form of communication by truckers and RV owners traveling the interstate, channel 9 is reserved for emergency communications only and is typically monitored by the state police, who have a separate CB radio mounted in their cruisers in addition to their other radios. Channel 19 is reserved for nonemergency news and updates concerning road conditions, warnings of traffic problems, and where "Smokey" is hiding.

During a grid-down event, a vehicle-style CB radio and separate 12-volt DC power supply can make an excellent "base station" for a rural home or off-grid cabin, paired with other family members and neighbors using handheld CB walkie-talkies. If you do decide to purchase a handheld CB radio paired with a CB base station, please note that most of the detachable antennas supplied with these walkie-talkies are the short, "rubber ducky" design.

While these are less likely to break off when backpacking or walking perimeters, they will significantly reduce transmitter range. I suggest ordering the optional telescopic CB antenna, which extends much longer and has the same quick-disconnect as the short flex antenna that comes with each walkie-talkie.

Since most handheld CB radios are supplied with 120-volt AC-powered chargers, make sure you purchase an optional charger that will operate from 12-volt DC. All vehicle and base station CB radios are designed to operate from a 12-volt DC power supply, but during a grid-down event you will need a separate RV/marine group 24- or 27-size battery to power these higher-wattage radios.

The private form of communications with the greatest range is ham radio. Depending on the license held by the operator, transmission power can be over 1000 watts, with communication across continents and oceans depending on time of day and frequency. While learning Morse code is no longer required, a written test is required, and most people will need to attend a weekly night class for several months to prepare, using a well-organized workbook. However, hams are very welcoming to all new members, and almost every city in the United States has ham radio volunteers teaching free classes several times each year. For information, try contacting the American Radio Emergency Service (ARES) at http://www.arrl.org/ares, or the Radio Amateur Civil Emergency Service (RACES) at http://www.usraces.org/.

TWO-METER BAND RADIO

The most basic level of ham radio today is the Technician Class license, which is the first step for anyone wanting to become a ham radio operator. This license limits which bands you are allowed to use, but does include the popular two-meter band.

Preppers have recently discovered with the basic Technician Class radio license they can use a small, two-meter FM handheld transceiver to communicate through a remote repeater station for crystal clear communication over a far greater distance than GMRS or CB radios.

Two-meter 75-watt ham transceiver

Although still a "line-of-sight" form of communication due to the very high frequency involved, there are two-meter repeater towers located at high elevations near most cities and towns. When any handheld two-meter radio is programmed with the correct "access codes" for a specific repeater, it will receive the weak transmission, then instantly rebroadcast the signal using its much more powerful transmitter and antenna system. While any two-meter radio can communicate directly with any other two-meter radio without using a repeater, the range drops to only a few miles, while repeater-based communication can cover a much larger area.

While the Technician Class license and two-meter radios may be the only level of emergency communications most preppers desire, there are additional levels of ham radio licensing that offer far more range and communication opportunities. The General Class amateur radio license requires a more thorough knowledge of radio and antenna theory, but is still a relatively easy next step for anyone having a Technician radio license.

The General amateur radio license allows operating in many more frequency "bands." Transmit power can be from 100 watts up to over 1000 watts depending on band and type of transmission, so any prepper with this license and higher level of communication equipment will be able to talk with other ham radio operators in every part of the world. Up-to-the-minute news from distant countries can be obtained firsthand without government filtering or watering down by official news outlets. The ham radio bands are set aside for individuals, so business use and advertising are strictly prohibited.

Even the lower-wattage ham transceivers can still communicate great distances and are designed for both base and mobile operations. A separate 120-volt AC power supply is required to provide 12 volts DC to power these radios during normal times, so powering directly from a deep-cycle RV/marine battery is easy when in a grid-down situation.

One minor word of warning: I have noticed some big game and bear hunters in remote parts of the country are starting to use marine-band walkie-talkies for communication. Marine-band radios typically have much more range and wattage than a CB radio. They do not require a repeater, like most two-meter communication, and the higher-frequency FM transmissions provide crystal clear reception. There are eighty-eight fixed channels assigned between 156 and 162.025 MHz, with channel 16 the international distress channel. Other channels are preassigned to harbor control, bridge control, international shipping, Coast Guard, and private over-water government channels.

I have been advised that several government land-based rescue agencies are starting to monitor channel 16 for hunters in distress. It's possible the FCC may look the other way if a bear hunter is being mauled by a bear and used his marine-band walkie-talkie to call for help on channel

16. However, it is still very, very illegal to use any type of marine-band radio over land, and if the user is caught, the fines can be severe.

If this is you, you'd better be in dire straits and not using these radios to call your hunting buddies back to camp for dinner or check on their location! It's also possible that automated satellite systems used for marine rescue work might be accidently triggered by someone operating on these marine frequencies when over land. You may suddenly have several black helicopters hovering over your campfire after homing in on your radio signal!

You may never have the desire to get into ham radio or attend classes and take license exams. However, as we close this chapter, please note that your emergency radio preparedness should include a quality battery-powered all-band radio receiver, and several FRS or CB walkie-talkies, since these radios do not require any type of license to operate.

As noted at the start of this chapter, having several different types of emergency radios and ways to keep them operating without grid power could save your life if a major disaster strikes your area. The next few chapters will deal with other emergency preparedness systems and equipment, and multiple ways to power from batteries.

8

COMPUTERS WITH BATTERY POWER

WE ALL WANT TO STAY CONNECTED with the outside world, especially during a grid-down situation. Regular broadcast news may be limited or controlled during extended emergency conditions, and being able to keep in touch using the Internet to access alternative news outlets and emergency information will be a real lifesaver, at least as long as the Internet is operational.

In my opinion, the greatest invention in home computers was the laptop computer and longer-life rechargeable batteries. No longer do you need a television-size monitor and desktop-size computer and

Charging laptops with solar power

keyboard requiring hundreds of watts to power. Recent advances in microelectronics and battery technology have now taken these already small laptop computers down to tablet and iPad size, while drastically increasing processor speed and battery life. Every laptop and tablet-size computer sold today includes either a built-in cell phone modem, wireless Internet modem, or high-speed wired Internet connection.

While cell phone and satellite Internet connections should still be available during the first few weeks of a grid-down event, direct-wired cable Internet and wireless Internet routers that may be needed to connect to the Internet will not work without grid power. Be sure the computer you plan to use during an extended power outage includes an alternative way to access the Internet, as wireless routers or home satellite modems are almost always 120 volts AC and grid powered.

During the initial stages of a grid-down event, you may be able to keep a laptop or tablet computer fully charged from a generator while it is powering other loads, or from the 12-volt utility outlet in your car, assuming you have a 12-volt adapter and connecting cable. However, at some point the generator will run out of fuel and you may need to save any remaining gas in the car or truck for a possible evacuation later.

During an extended power outage, the easiest way to keep a laptop or tablet computer charged without a generator is with a fold-up solar module, which is available in 5-, 10-, 25-, and 40-watt or higher capacities. The 5- or 10-watt models will easily recharge your cell phone. You will need a minimum of 25 watts to recharge most tablet computers in a reasonable time, and up to 40 watts for larger laptop computers.

Even a small 10-watt solar charger can eventually recharge a larger laptop computer. However, it may take days of solar charging to provide one or two hours of computer use, so make sure you have at least one of the higher-wattage modules to recharge any of the larger battery-powered devices.

Although fold-up solar modules over 25 watts can be fairly expensive, they will significantly reduce charging time, which is a real advantage on winter days, which have fewer hours of sunlight. A fold-up solar

COMPUTERS WITH BATTERY POWER

40-watt fold-up solar charger powering laptop computer

charger is a must-have item and can be easily packed and taken with you if you do need to bug out. Some of the newest fold-up solar chargers are available with a built-in high-capacity battery. This allows charging the internal battery during the day when placed in the sun, which can then be taken inside to recharge multiple small electronic devices at night or during periods of limited sunlight.

A battery-powered laptop or tablet computer with a built-in wireless Internet or cell phone modem is a great way to keep in touch with friends and family by providing texting, e-mail, and public forum postings. Portable computing devices provide the ability to do an Internet search on any topic, read breaking news articles, obtain movie listings, access street maps, and get travel directions by typing just a few words in a search engine. You can instantly download radio show podcasts and instructional videos and scan late-breaking headlines from multiple news sources.

However, even if you are operating on battery power and have multiple ways to keep the laptop or tablet computer fully charged, what happens if there is no Internet available to access this information?

Start collecting reference manuals and anthologies on CD.

A local power outage can take down wireless Internet routers, cell phone towers, and satellite modems. In addition, a full grid-down event may cause communication blackouts over a much wider area and lasting months. Then what?

While things are still relatively normal, this is a perfect time to download reference books, self-help articles, magazine archives, repair and parts manuals, code books, emergency medical procedures, pill and medical dictionaries, address lists, and anything else you regularly reference for your work or hobby. Most magazine articles can be downloaded and saved on your computer for reference later, and most publishers offer multiple-year anthologies of their magazines on CD in an easy-to-search format. There are thousands of "classic" books available to download for free due to expired copyrights.

Several publishers specializing in prepper products have copied thousands of pages of survival and emergency medical articles onto easily searchable memory sticks for access without the Internet when plugged into your computer's USB port. I suggest purchasing a quality encyclopedia in electronic format on CD or as a downloadable app. While not free, their cost is minimal and they will be a great way to research almost any topic without Internet access or taking up several feet of bookshelves that you can't take with you if you do need to evacuate.

All mapping programs require Internet access, but it's still possible to buy a good map database that can be totally downloaded and will work with or without a GPS receiver or Internet connection. There are many types to choose from based on geographic features. The cost is substantially less if you need only a detailed street or terrain map for a specific state instead of the entire country.

Some mapping programs for cell phones will provide exact GPS location and directions while walking along city streets, biking, or even hiking on an unknown mountain trail. However, many software packages download only the map section for your current location to save memory and may require reestablishing an Internet connection to download the map information for the area you are traveling into. Make sure the map software you purchase clearly indicates it will work totally off-line using only the map database that resides on your computer device.

If you do have to bug out and are traveling through areas with no cell or satellite Internet service, being able to review a built-in map database is a necessity, especially if you cannot access a foldout paper map. Preplanning your bugout travel should identify routes that avoid bridges, interstate exit points, and congested cities that could become a major traffic bottleneck or checkpoint during a crisis that would bring normal traffic to a complete standstill.

Finally, while you are downloading and storing all this reference material onto your battery-powered computer, it's also a good idea to scan important documents that could be destroyed in a fire or flood.

Homeowners and car insurance documentation, property deeds, financial and medical records, and banking and credit card information can all be safely stored using file encryption if your computer is lost or stolen. Irreplaceable family photos can also be saved and reprinted later if the originals are destroyed. Many big-box stores offer do-it-yourself printing of photographs from a CD disk or memory stick, and the quality of these color photographs is equal to the original.

While having a battery-powered laptop or tablet computer is critical during a major power outage, it will be even more useful if you can still access all of your important documents, maps, and reference materials when all communication systems are down.

I encourage scanning copies into your laptop computer of all important documents, insurance policies, vehicle titles, property deeds, banking information, and family photos just in case you do have to evacuate. However, this should not be your only record of these documents. We have an accordion-style expanding file folder that is compact and designed to be carried. We keep this up-to-date and nearby in case we have to evacuate. It has hard copies of the same information scanned into the laptop since it's possible this information would not be available if it is lost, has a failed hard drive, or is stolen.

9

ENTERTAINMENT WITH BATTERY POWER

DURING AN EXTENDED POWER OUTAGE AND long after the grid and your generator have died, you will not have the power to operate large-screen televisions, CD music systems, DVD movie players, and Internet-based entertainment devices that run on 120-volt AC power. In addition, if this is a true grid-down event, all Internet and normal communication systems will have stopped, even if you can maintain emergency power for your electronic equipment.

TV AND DVD PLAYERS

Thanks to the recreational vehicle industry and all the parents needing a way to keep the kids in the backseat quiet on a long drive, there are now many sizes of battery-powered flat-screen televisions and DVD players available. These are supplied with both a 120-volt AC charger and a cigar-type plug to fit a 12-volt DC vehicle utility outlet. Assuming broadcast television stations are still on the air, these portable digital televisions can receive all broadcast stations in range. They also come with a short antenna having a magnetic base and long cable, allowing attachment to your vehicle roof for mobile use. If you live in a rural area, I recommend buying an extended-range digital broadcast antenna with signal booster. Although this requires 12-volt DC power, it will greatly extend the range and number of channels your battery-powered television can receive.

12-volt battery-powered DVD movie players

Battery-powered DVD video players are also popular for backseat kids entertainment. A DVD video player does not include a broadcast television tuner, but its built-in DVD drive and fold-up color screen with stereo speakers are great for one or two people to watch a favorite movie when other forms of entertainment are without power. I recommend stopping by the discount DVD bin each week to build up a good supply of movies you can watch off-line, especially if you have young kids. I purchased several multiple-sleeve DVD holders after discarding their plastic cases, and I can fit several hundred DVDs into each compact case that will be easy to grab along with your player if you do need to evacuate.

Most of the portable DVD video players intended for vehicle or camping use have fairly small screens, but there are much larger flat-screen digital televisions sold for the mobile RV and boating industry. These are especially nice during a power outage, as they are designed to run on 12-volt DC power. This means they are not only more rugged than a stationary home television, but they are also designed to be very energy efficient since they are battery powered.

There are several brands of dual-voltage flat-screen televisions with removable bases and wide-screen format from thirteen inches up to twenty-two inches in size. In addition to a broadcast television tuner, most include a built-in DVD player and input slots for playing videos or showing photos from SD cards and USB-connected sources. The screen on the Axess 15-inch model I purchased was easily viewable by the whole family, and the built-in DVD player and stereo speakers only required connection to a 12-volt DC power source. Since battery life was my main concern, when checking battery drain, I found the 15-inch model used half the power of the 22-inch model.

If planning to bug out alone to a remote cabin, the smaller 12-volt DC personal-size television with DVD player will be the easiest to pack and requires the least amount of battery power. However, for two or more people I recommend the fifteen or nineteen-inch size dual-voltage flat-screen televisions having a built-in DVD player and stereo speakers.

12-volt DC-powered 15-inch TV with DVD player

If you are not a recreational vehicle owner, I strongly suggest checking out your nearest RV supply outlet. Remember: with the exception of the microwave oven and rooftop AC unit, every LED light fixture, kitchen appliance, water pump, and entertainment device in any size RV is powered by a 12-volt deep-cycle battery. In addition, since audio/video equipment designed for RVs and boats is portable and battery powered, it is designed to be very rugged with a minimum battery drain. You can find the larger flat-screen televisions at any RV supply store, as well as 12-volt DC-powered fold-up portable satellite

iPad with 12-volt DC screen projector

dishes for both Internet and cable television reception anywhere.

Laptop computers and iPads are not only battery powered, but they can provide hours of entertainment through computer games, videos, and family photography.

MUSICAL INSTRUMENTS

Most people at first would not consider live music to be a major priority during a real grid-down period of weeks or months without electricity. However, being able to host an impromptu jam session, lead an in-home church service, or amplify a speaker's voice at a gathering of neighbors could bring a little comfort and entertainment during an otherwise fairly bleak period.

Some home-model digital pianos and portable organs have a built-in amplifier and stereo speakers and require very little power. Many actually operate on 12-volt DC and use a power supply in their plug-in cable to convert from 120-volt AC grid power. This means these models are already able to operate directly from an external battery if you have the right power connector. In addition, there are several portable high-quality PA amplifiers available that include internal 12-volt batteries and allow operating from either grid power or battery power.

If you do not need a full-size keyboard, there are several keyboards that have different adaptor cables that will plug into an iPad or laptop computer. These obtain their power directly from the computer port and do not require batteries or an external power supply. Depending on the software, these can generate studio-quality piano or organ sounds, as well as hundreds of other instrument sounds.

I am not suggesting that you run out to buy a battery-powered keyboard or guitar amplifier and put it in storage for some future power outage. However, when shopping for a digital keyboard or compact amplifier, check to see which brands and models are capable of dual-voltage operation, as there may come a time when this capability will be much appreciated.

Battery-powered PA amplifier

We do not normally think of providing entertainment during a short-term power outage since trying to provide emergency lighting and keeping everything from melting in the refrigerator are the primary concerns. However, when a power outage extends beyond a few days, and some areas taking weeks to have all power restored, your family needs opportunities to take their minds off what they are going through, if only for a few hours.

Being able to utilize battery-powered CD music systems, DVD video systems, computer games, and musical instruments can go a long way toward calming nerves and fighting boredom, especially for younger members of a family. However, sitting by candlelight and playing card games or checkers or working on a puzzle require absolutely no electrical power and can be very entertaining. These are low-cost entertainment items that you should have stored away for emergencies, and they fit nicely under the couch.

10

MEDICAL EQUIPMENT WITH BATTERY POWER

WHILE KEEPING THE LIGHTS AND A television operating during an extended power outage makes life easier, keeping medical devices powered up and medicine refrigerated can be a matter of life and death during a grid-down situation. I have designed emergency backup systems for critical care patients living at home, and some required an entire roomful of electronic monitors, pumps, and refrigeration equipment.

Providing long-term backup power for a family member needing this level of home care is beyond the scope of this text. However, keeping a CPAP (continuous positive airway pressure) or BiPAP (bilevel positive airway pressure) ventilator unit or heart monitor operating through the night is fairly simple, as these small loads can be powered by their own battery backup system.

Most bedside medical monitors and ventilating units not only require very little electrical power, but they can operate from optional

MEDICAL EQUIPMENT WITH BATTERY POWER

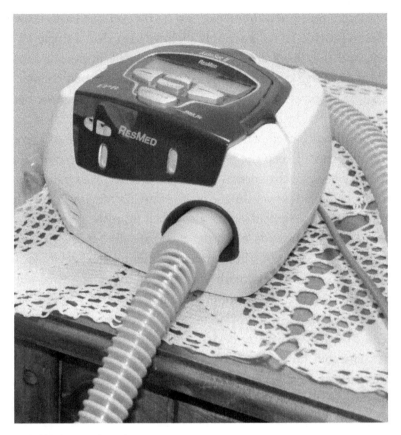

12-Volt DC-powered ventilator unit

12-volt or 24-volt DC emergency power supplies. A new BiPAP unit I recently tested during actual use consumed only 0.10 kWh of power at 120-volt AC (9 amp-hours at 12-volt DC) for a typical thirteen-hour runtime per night. This would indicate that a BCI group 27 size 12-volt RV/marine battery should easily power this unit for five days without totally discharging the battery.

The most reliable solution is a totally separate battery-backup system to power this medical equipment. A sealed 12-volt gel or AGM battery can be located near the medical equipment since there is no battery outgassing from these sealed batteries under normal operating conditions.

For larger power requirements, two 6-volt golf cart batteries wired in series to provide 12 volts are only slightly larger than a single 12-volt battery, and they have a much higher combined amp-hour capacity and deep discharge cycling ability. Having a separate solar charger and battery just for powering medical devices will avoid having the battery drained by noncritical electric loads.

This isolated battery can be connected to a 120-volt AC "trickle" charger, which will keep the battery fully charged during normal grid operation. The same charger can also be powered using a small generator during an extended power outage by just moving the power cord from the wall outlet to an extension cord supplied from the generator.

Some wireless bedside monitors automatically report back to the medical staff the status of a heart pacer or other medical device at night using a phone line. Check with your doctor to see if a dual-voltage model is available that allows powering from a 12- or 24-volt DC battery supply. If the grid goes down, you may not have a working generator to power 120-volt AC medical devices, especially if they must stay powered twenty-four hours per day, seven days per week. This would be a significant load on any backup power system and could quickly drain the fuel supply for any generator.

A quick, temporary solution to keep a CPAP machine working during a power outage, especially one that occurs unexpectedly during the night, is to keep it plugged into a larger-capacity uninterruptable power supply (UPS). These are popular to keep a desktop computer from crashing if the power goes out right in the middle of an operation or when copying files. Since any computer system not powered from batteries will shut down the instant there is any problem with the electric grid, these UPS devices are designed to keep the power flowing when there is any power interruption or quality issue with the grid.

These units are left plugged in and the grid power is routed straight through to their power outlets, which in this case would be a CPAP machine or other medical devices. During normal conditions a small part of this grid power is used to keep the internal battery fully charged and on standby. A fraction of a second after the grid goes down, the UPS system will switch over to the 120-volt AC inverter circuit powered from the internal battery and provide normal power to any connected appliance.

Of course, these systems cannot power a medical device indefinitely, as the normal runtime for most connected desktop computers will be measured in minutes, not hours or days after the power goes out. However, most CPAP machines and medical monitors will have a much smaller drain on the UPS battery than the combined load of a desktop computer, monitor, and printer.

UPS systems for desktop computers are sized based on the total "watts" of connected load they can power, and their volt-amp (VA) rating, which determines how long they can power a given load. Typically, a medium-sized UPS unit will have a capacity rating around 900 VA, or about 0.9 kWh. If your CPAP machine requires 0.2 kilowatt-hours of electricity per night, we would expect this medium-size UPS system to provide about four nights of operation.

Of course, actual time will vary, and some UPS units will automatically turn off before the battery is completely drained to protect the battery. We are not too concerned with the "watt" rating when selecting a UPS unit since almost any CPAP or medical monitor will draw far less load than any computer these UPS units are designed to power. Ideally you would use the UPS system together with a small backup generator during an extended power outage and would need to run the generator only a few hours every few days to quickly recharge the UPS system's internal battery.

Chapter 13 covers how to keep a DC refrigerator and freezer powered during a grid-down event, and this is even more critical if you have medications that must be refrigerated.

Chapter 21 describes how to build your own emergency solar-charged battery backup system that will keep your CPAP machine operating indefinitely without the electric grid. A small, 10-amp solar charge controller and a 40- to 50-watt solar module are a good match for the daily load of these medical devices when both the electric grid and generator are no longer working.

Some people will have major health problems if they go even one day without the use of their grid-powered medical devices. However, power outages lasting weeks are occurring somewhere in this country every day and becoming more common as the utility infrastructure ages. Not having a backup plan to keep these medical devices operating during an extended power outage could be life-threatening for you or members of your family.

The information in the next few chapters will provide additional ways to power these medical devices. There have been some major improvements to make these devices smaller and requiring much less electrical power to operate. In addition, many manufacturers now offer backup battery packs to power their more efficient models, and an upgrade may make alternative charging methods much easier and less costly.

11

PORTABLE TOOLS WITH BATTERY POWER

DURING A LONG-TERM POWER OUTAGE, IT is doubtful you will need to build a house. However, you may need to repair storm damage to an existing house, board up windows, or repair a vehicle. You may have a garage full of power tools, an air compressor and pneumatic tools, and a large selection of electric saws and drills. However, these will be useless during a true grid-down event, especially after your generator has run out of fuel.

I strongly recommend buying a complete set of portable, battery-powered tools that use the same size and type battery. There are several name-brand suppliers offering cordless power saws, drills, work lights, and cut-off saws, and all use the same rechargeable batteries. If funds are limited, you can't go wrong with a battery-powered reciprocating saw. These saws are amazingly versatile with different specialty blades for cutting wood, metal, and even wood having embedded nails for demolition work after a storm.

Battery-Powered Tools

I served several years with a local volunteer rescue squad, and we had a reciprocating saw on every rescue vehicle and fire truck. This should be your first battery-powered saw to start your collection, and those based around an 20-volt battery are a good choice for both cutting power and longer battery life.

Your second battery-powered tool should be a multispeed one-half-inch-capacity drill. In addition to all sizes of drill bits, there are screwdriver blades, wire brushes, and even paint stirrers to fit these drills. There are also small water pumps available to fit a battery-powered drill, which can be connected to a garden hose for both filling storage containers and removing standing storm water.

Since many manufacturers are converting over to longer runtime lithium-ion batteries for their cordless tools, they now even offer battery-powered chain saws, which actually work reasonably well with these higher-capacity batteries.

A battery-powered chain saw stored in your truck can be a real lifesaver for clearing downed trees and limbs blocking your evacuation route. Since it requires no gas, it can easily store behind the seat.

Most brands and sizes of chain saws, regardless of being gasoline or battery powered, tend to leak bar oil during storage, which can ruin vehicle carpets or upholstery. To avoid this mess you may want to leave the automatic oiler tank empty and instead apply chain oil using a separate squirt bottle to spray oil directly onto the chain when cutting.

12-volt charger to charge 18-volt power tools

While any battery tool purchase usually comes with its own 120-volt AC battery charger, without grid or generator power you will have no way to recharge these power tools once they are discharged.

While not normally found in a typical builder-supply outlet, battery-powered tool manufacturers also offer a separate charger that plugs into your car's 12-volt DC utility outlet and does not require the grid or a generator to operate. Battery-powered tool chargers that plug into a 12-volt vehicle outlet instead of a 120-volt AC wall outlet are typically only available from the manufacturer of your battery-powered tools. Most will let you order directly from their website. It is very important to have several battery-powered tools and a way to keep them charged during a grid-down event. In addition, since these special chargers are designed to operate from a 12-volt DC vehicle outlet, they can also be powered from a separate deep-cycle 12-volt battery that is charged with a solar module.

Keep in mind the importance of standardizing on a single battery brand and size so you can use a common AC charger and DC charger to keep them all operating during both normal, everyday times and grid-down times. Another advantage of standardizing on a single brand is you will not need a separate size or type of battery for each power tool.

An AGM battery can be used to make a homemade charger to keep your battery-powered tools charged when grid power is not available. It consists of a 12-volt deep-cycle sealed group 27-size battery and battery box, wired to a standard 12-volt vehicle utility outlet. The wiring diagram in the appendix provides more detail on this unit, and includes wiring information to add a solar charging option.

Having a primary battery and one or two spare batteries that can be charged while a tool is in use makes it easy to always have several fully charged batteries that can be shared with whatever tool is needed at a given time. This also avoids the problem of having a battery staying in each separate tool and they all are totally discharged due to minimal use of each tool. Not having a battery for every tool may save enough to buy the manufacturer's optional higher-capacity batteries and a faster charger. During emergency conditions, this will allow both longer tool

AGM battery and 12-volt charger for off-grid tool use

operating times and faster battery recharge times.

While you do not want to overload your source of backup power, most brands of chargers for battery-powered tools offer both standard and fast chargers. Although the fast chargers draws more amps, in some cases they can cut your charge time in half. This is a real advantage if you are powering from a generator, as this could also cut fuel usage by half.

Many extended power outages are storm related which often includes wind and water damage to roofs, windows, and vehicles. Having battery-powered drills and saws that operate during a power outage can make temporary repairs much easier and quicker. Being able to board over broken windows and patch a leaking roof until major repairs can be completed can prevent further water damage to interiors. If you do have to bug out, power tools can make temporary shelter construction much less labor intensive. Finally, when it comes to making temporary repairs, don't forget the duct tape.

12

CLEAN WATER WITH BATTERY POWER

THE HUMAN BODY CAN SURVIVE UP to several weeks without food, depending on the amount of stored body fat. However, we cannot live longer than two or three days without water to drink. In any survival or emergency situation, securing an adequate supply of fresh drinking water for each member of your family is your primary preparedness concern. After any disaster that interrupts the normal supply of fresh drinking water, outbreaks of widespread illnesses soon follow as a result of drinking contaminated water.

You need a simple and foolproof backup plan to have clean water, even if this means no running water in the house. For thousands of years civilizations relied on lifting or pumping well water to the surface by hand or animal power, and carrying it where needed using buckets or clay pots. Today we rely on pumps supplying city water or well pumps supplying well water but these do not work without utility power.

WELL WATER

There are several easy ways to obtain clean water from an existing well when the grid is down. My two-hundred-foot-deep drilled well has a standard 240-volt AC well pump located near the bottom, which is piped into the house. It is normally powered by the utility grid, but if the grid is down, it can be powered by my whole-house generator during a power outage. I also have a 24-volt DC well pump located in the same well, but held slightly above the AC pump. Both pumps are piped together using check valves at the point where the separate supply pipes exit the well casing. This avoids one pump trying to pump back into the piping of the other pump.

My pressure tank has two pressure switches, with the highest pressure setting used for the AC pump and the lowest setting for the DC pump. If both grid power and generator power are lost, the AC pump will stop pumping and the water pressure will continue to drop until the switch controlling the DC pump is energized. Although the DC pump is much smaller and pumps less than half the flow and pressure of the AC pump, it is always available since it is powered directly from my solar-charged 24-volt battery bank and does not require an inverter to work.

12-volt DC well pump

I also installed the largest pressure tank I could find, which is about the size of a hot-water tank. The large storage capacity significantly reduces the constant cycling of the pump. Constantly starting and stopping any pump is hard on both the pump and an inverter or generator powering the pump, and will shorten their operating life.

If your well is not close to the house or you have a different well located at some distance from your home or weekend cabin, another solution may be to use a 12- or 24-volt submersible pump as the only pump in this well. It will be powered from a nearby pole-mounted solar array which keeps the solar module(s) and wiring up and away from potential animal damage.

This system does not require any batteries, and the operation is very basic. This is also a very common method used on today's larger farms and ranches for watering cattle in distant fields where there is little surface water. These solar pumping systems are also sometimes used to keep cattle from fouling small streams if they did not have this other source of drinking water.

When the sun is up, the pump starts pumping, and when the sun goes down, the pump stops pumping. Of course, you will need someplace for this water to go, as even a small, three-gallon-per-minute solar pump can pump over twelve hundred gallons in a single day. Typically, this water is piped directly into a concrete or galvanized-steel storage tank, located at a higher elevation than the house or cabin. For every 2.3 feet of elevation, you will get an additional 1 psi of water pressure, so to achieve a typical 30 to 40 psi city water pressure, the tank will need to be located seventy to ninety feet higher.

Keep in mind this is a change in elevation, not a measurement of the distance between your home and the tank. In northern climates, to prevent freezing, the tank may need to be below ground unless it is several thousand gallons in size and the water flow out is fairly constant.

Solar array to power nearby 12-volt pump

You can wire the pump directly to the solar module without a solar charge controller or batteries and it will work. However, installations that include an optional "pump controller," will improve the pumping and protect the pump. During the first and last hours of daylight, the solar array may still be producing voltage, but not enough amp flow to start the pump or to keep it pumping. Without the controller, the pump will stall and stop pumping but will still have some electrical power passing through the pump's motor. Over time, this can cause the motor's windings to overheat and eventually short out and even dry out the lubrication in the motor bearings.

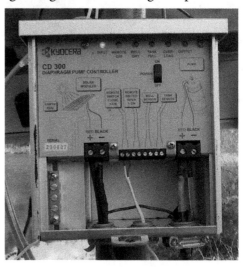

The pump controller blocks power from going to the pump until it is within the voltage and current range to run safely. In addition, more expensive pump controllers can convert any excess voltage to additional amp current during these lower sun periods, which can power the pump when it would normally not have enough current to run.

Pump controller for solar-powered well pump

Most pump controllers also include replaceable fuses to protect the wiring and added terminals that allow connecting float switches. These typically look like a rubber softball and contain a waterproof switch and several feet of waterproof cable. These can be installed in the tank to start and stop the pump based on water level in the tank, and can also be installed in the well to stop the pump if the water level gets too low. While this may sound a little complicated, the low voltage is reasonably safe, and prepackaged kits are available that will include everything you will need but the pipe.

Modern hand pump for existing well casing

For those seeking a less technical solution, there are several high-quality hand pumps now available that fit on the top of a standard four- or six-inch well casing. These are not the old cast-iron pitcher pumps with leather plungers you see in photos of farmhouses during the 1930s. These have all stainless steel and brass construction, and will provide many years of service. These new hand pumps are easy to install since their PVC dip tube and check valve easily extend down into the well past any existing piping. The hand pump's base replaces the existing well cap and is attached by bolts. Most models have an optional 12-volt DC motor drive that can be powered during an extended power outage using your vehicle's battery or a separate solar array.

TOILETS

Without grid power, once the generator is out of fuel, there may be no running water in the house unless you have a solar backup system. This also means no functioning toilets unless a nearby pond or creek provides flushing by water bucket. However, a composting toilet is a great backup plan, as it requires no sewer connection and no water plumbing, so it can be located anywhere.

These compact units do not require any water to flush, and most

CLEAN WATER WITH BATTERY POWER

models have a small, 12-volt DC exhaust fan that constantly pulls air from the toilet and exhausts to the outside, eliminating all indoor odor. A small scoop of mulch dumped into the manually rotated drum is all that is needed after each use, and a bag of mulch will last several months.

WATER FILTERS

During a grid-down event, clean drinking water may not be available if you are on a municipal water system. This is even more of a concern if you live in a densely populated city or an apartment having limited storage space. Clean drinking water is also a concern if you must evacuate or bug out with only what you can carry. We purchased a large Berkey water filter for our home, and a smaller model for our truck camper.

Composting toilet with 12-volt exhaust fan

While the replaceable filter elements in each water filter last much longer if you are not trying to filter cloudy or sediment-filled surface water, these filters will still remove all contaminants, including bacteria, parasites, cysts, lead, and mercury. In addition, suppliers now offer both "straw" type, individual-user water

Berkey water filter in RV camper

filters and portable models having battery-powered ultraviolet LED lamps that kill all pathogens.

Being able to provide your family's drinking water each day from virtually any surface-water source means you will not be standing in line for hours with everyone else waiting on a FEMA truck to hand out a few bottles of drinking water.

My ninety-four-year-old mother lives near Charleston, West Virginia. On January 10, 2014, a chemical storage tank located just upstream from the Charleston water plant on the Elk River developed a large leak of a strong "foaming agent" used to clean coal. (Why any government agency would allow a chemical company to locate a chemical tank farm just a half mile upstream from the main water intake of the water treatment plant serving nine counties, the state capitol, and three hundred thousand residents is another story.) While 4-methylcyclohexanemethanol isn't lethal, drinking water with even a small trace causes vomiting, difficulty breathing, and blistering of any skin area it contacts.

The biggest problem in trying to get this chemical out of the city water supply was that "foaming agents" are actually used in water treatment plants to help remove many types of contaminants, and there was no foaming agent designed to remove another foaming agent! The result was, the chemical stayed in hundreds of miles of water pipes for weeks, and residents were warned for months not to drink or cook using this water.

As soon as this hit the national news around noon, I loaded up a 200-gallon water storage tank I have that fits the bed of my pickup and called to warn her not to drink, wash her hands, brush her teeth, or bathe using the contaminated water. I live in Virginia, about four hours away, and when I arrived that evening, I found a major city in crisis. Every hotel and restaurant in this large metropolitan city was forced to close immediately. Later that night the state police were called in to protect the night crew of a local Wal-Mart as they tried to unload a tractor-trailer load of bottled water, which was surrounded by angry residents with absolutely no water to drink and in no mood to wait for the store to open the next day.

The following morning, after filling multiple gallon jugs and a thirty-gallon water storage container I left behind full of fresh well water, I still had almost half a tank on the back of my truck, so I headed down the street, asking anyone I could find outside if they would like free water. I could not believe how many people not only do not keep any water backup, but didn't even own a water jug I could fill. People were coming out to me with empty kitchen trash cans and mop buckets. One man came out with a boxful of small plastic soda bottles and asked me to fill them individually.

I have returned many times to Charleston since then, and even two years later most restaurants still have signs posted that indicate no city water was being used for cooking or washing. I can assure you nobody in those nine counties expected this to happen, and even though everyone still had power, heat, and phone service, city residents suddenly and without any warning found they could not wash, drink, eat, or bathe, and most had absolutely no emergency water supply.

Most county fire departments started using their tankers to truck in water from distant fire stations not impacted by the chemical spill, and each became a central distribution point. Unfortunately, many elderly and disabled residents had no way to drive to these distant locations. When FEMA finally arrived days later with trucks of bottled water, there was no distribution set up to reach these elderly residents. In addition, while these small bottles provided safe drinking water, there was still no realistic solution to resolve the washing and bathing problem.

One of the first things that happens in any disaster is the local population have no stored drinking water and soon become sick from drinking contaminated water. A tabletop water purifier at home, several gallons of bottled water in storage, and a portable water filter in a bugout bag are must-haves. Don't go another week without them.

13

REFRIGERATION WITH BATTERY POWER

REFRIGERATORS AND FREEZERS DO NOT REQUIRE continuous power, but since they cycle on for a few minutes every hour depending on their insulation thickness and the room temperature, power needs to be available every time they do cycle on to keep things cold. Tests have shown that constantly opening and closing the doors of a partially filled refrigerator or freezer will significantly increase compressor runtime and energy use.

Most new refrigerators are better insulated and should be able to hold their temperature through the night if their doors are kept closed, but they will need to cycle on again first thing in the morning to make up for any temperature increase during the night. Early morning is the time family members will be getting up and wanting to bathe and eat even during a grid-down event.

Manually starting a generator during the early morning will minimize generator fuel use by powering the well pump, cooling down a refrigerator, and running the coffeepot, microwave oven, and a few lights since these will need to operate at the same time you are preparing breakfast. Running a whole-house generator for two hours in the morning should provide enough runtime to cool down the refrigerator and freezer while everyone is getting up, bathing, preparing breakfast, and eating.

After breakfast the sun should be up and providing plenty of day lighting in most interior areas of the home or cabin, and the refrigerator should be cooled back down enough to coast until evening if the door is kept closed. Manually starting the generator again for two hours during evening meal preparation, when the well pump, room lights, microwave oven, and other kitchen appliances are needed again, will also allow the refrigerator time to cool back down.

By concentrating electrical power needs to the periods of breakfast and dinner, your generator runtime will be kept to a minimum while still keeping the refrigerator reasonably cold. During a real grid-down event, supper may become the main prepared meal, with breakfast and lunch consisting of packaged snacks or foods that do not require cooking.

I recommend using an external temperature display for your refrigerator, which will help determine if the refrigerator is warming up without constantly opening the door to check. Gasoline and propane are too valuable to waste running a generator throughout the day during an extended power outage, especially just to power a few lights or small appliances that could be battery powered.

While most total off-grid solar homes will have enough solar power to operate a conventional 120-volt AC refrigerator, this is still one of the largest electrical loads in any home and may become problematic in a grid-down situation. Buying a slightly smaller refrigerator with a high efficiency rating is a great first step, but you will still be dead in the water if your generator or solar system fails and cannot keep the backup battery charged.

If an extended power outage lasts longer than three or four weeks, you will probably have already eaten everything in your freezer anyway, even if you have some alternate way to keep it operating. In addition, if your entire area is still without power, any gas station or grocery store within a reasonable travel distance will also be dark and empty.

With nothing to refrigerate, I am afraid trying to power a conventional refrigerator or freezer at this point will be a waste of effort and scarce resources. History has shown us time and again that meals and meal preparation will become a much simpler affair after all of your kitchen appliances stop working, the store shelves are empty, and there is no water coming out of the faucet to use in preparing meals and washing dishes.

The primary meal during a true grid-down event will be a large pot of soup with an attempt each day to provide some variety. No more thawing a frozen chicken under running water, turning on the oven, and setting the table. However, while there may be limited need for a refrigerator, being able to cool a drink on a hot summer day with a handful of ice cubes may become one of the few pleasures left long after the grid fails.

I strongly suggest owning a top-load, 12-volt DC-powered freezer as the main refrigeration for an off-grid cabin, and as backup refrigeration for a conventional kitchen refrigerator in a grid-connected home. Keeping meats and ice frozen using a solar-charged battery means you will not need to host a backyard barbecue right after the grid fails and all the meats in the freezer start to thaw.

Having a DC freezer powered by a solar-charged battery is about as simple and fail-safe as you can get. If you do have to stay in a nonsolar home during a grid-down event, you can still benefit from having a DC-powered freezer. Most of the 12-volt DC powered refrigerators and freezers I have used will easily stay cold for days when powered by two deep-cycle 6-volt golf cart batteries.

This means you can keep these batteries charged using a grid-connected charger to offset the daily battery discharge of the freezer. You can also power the same battery charger by running a generator a few hours each day during an extended power outage, which will significantly extend fuel supplies. Finally, you can install a separate solar module and solar charge controller to keep the same batteries charged during a grid-down event without installing a whole-house solar power system costing thousands of dollars.

A top-load DC refrigerator in the smaller two- to three-cubic-foot size range will require an 80-watt solar module for full-time use in most parts of the country. A top-load DC freezer the same size will require a 150-watt, or larger, solar module for most parts of the country. You most likely will need to double these estimates for a five- to six-cubic-foot unit. The

A 12-volt DC Sundanzer top-load freezer

solar module needs to be sized to put into the battery (or batteries) each day a charge that is equal to the combined daytime and nighttime load on the battery. Using this sizing guide, the battery is never discharged more than 50 percent, which will greatly extend battery life.

While you will not find a battery-powered 12- or 24-volt DC refrigerator or freezer at your local appliance store, there are several very good quality brands available on the Internet. These include SunDanzer, Sun Frost, and Unique off-grid appliances.

While today's battery-powered inverters are fairly reliable, any conventional 120-volt AC refrigerator or freezer will still shut down if there is a problem with the inverter. Having a top-load 12-volt DC freezer means as long as there is any charge left in the batteries, and regardless of charging method, the freezer will still stay cold since no other electrical equipment is required. Being able to make ice also allows moving this ice to a conventional kitchen refrigerator or portable ice chest each day during a grid-down event. This procedure can keep foods, drinks, or medicine cold for as long as your supplies last.

If you do need to bug out or need to severely lower your electrical usage and stay at home during a grid-down situation, there are some new portable ice chests that are amazing. While even the smallest sizes are still expensive, their superefficient design and thick-wall insulation are worth the cost. The tight-fitting lid includes dual hold-down straps and thick door seals that totally eliminate air infiltration. These ice chests can keep ice and meats cold for days of bugout vehicle travel and camping, and are another must-have item.

Not everyone will have the extra space or budget to install a battery-powered freezer and solar charger to provide ice and keep foods frozen during a grid-down event. However, when all other grid-powered refrigerators and freezers have failed, having one of the new superinsulated ice chests can keep ice and meats frozen for several days depending on ambient temperature and how full the cooler is packed.

During a power outage expected to last only a day, it may be better to just leave everything in the kitchen freezer section and just keep the door tightly closed if you have a fairly new refrigerator with good door seals. However, if you are facing a power outage expected to last weeks before power can be restored, moving ice and frozen meats into one of the new superinsulated ice chests can provide several additional days of safe cold storage. This also is very helpful if you do have to evacuate. The next chapter will address areas where battery power can make life much less stressful and safer during an extended power outage.

14

SECURITY WITH BATTERY POWER

WITH THE RECENT ADVANCES IN WIRELESS security cameras and motion sensor technology, there are few homes or businesses today without at least some basic level of security. While older security monitoring technology still requires a phone landline to report a fire or break-in, newer systems can utilize cell phone and Internet modems to send alarms to a distant emergency call center. Unfortunately, while all security systems do include battery backup, most will fail when a power outage lasts longer than a week or two. In addition, during a true grid-down event there may not be a local cellular or Internet communication network still functioning that could report an alarm, assuming your security system is still operational.

Once a power outage extends beyond a few weeks, vandalism and theft will increase dramatically since the majority of any remaining residents still in any city or town will be without food and water, and local law enforcement will have far more to deal with than responding to a burglar alarm. If you feel a gun is the best burglar alarm, keep in mind that you have to sleep sometime, and some burglaries occur late at night when the homeowner is home and asleep.

Increased vandalism during a power outage is also why many portable generators tend to walk away from garages in the middle of the night. Even a professionally installed security system and the communications used to report alarms will fail sooner or later during a grid-down event. We need a much simpler way to notify us when somebody is trying to break in or steal our property during an extended power outage and all grid-powered alarm systems have failed.

ALARMS

While there are all kinds of simple security alarm tricks, including balancing a pop bottle on a doorknob or tin cans hanging from a string stretched across an outside walkway, there are portable battery-powered security devices that require minimum power and do not need any external communication connection. Primarily marketed for apartment dwellers and hotel stays when traveling, there are all sizes and types of portable door and motion sensor alarms that are self-contained and battery powered.

Battery-powered door alarm

Avoid the smaller-size models powered by "button" or odd-size batteries, which will be impossible to replace or recharge during a grid-down event. However, with a little extra effort you can find models that operate on either AA or C batteries. Chapter 4 explains how to keep these charged.

Portable motion alarms are the size of a cell phone and include a metal loop that hangs on any doorknob. Once the switch is activated, these devices provide an earsplitting alarm if the knob is turned or the door is disturbed. Some models allow adjusting the volume and switching between a siren or chime sound.

Another type of battery-powered alarm that can be used to guard a nearby garage or generator shed is a self-contained motion sensor. While very similar in appearance to your typical wall-mounted motion sensors connected to a central alarm system, these units are self-contained and can be easily moved between multiple locations. These motion sensors have a short delay when activated, allowing the homeowner time to move out of the sensor's range. Instead of an internal alarm, some models can activate a remote alarm unit using wireless technology.

As mentioned previously, make sure to select a model that uses either AA or C batteries, which are the same size batteries used in all other battery-powered devices reviewed in this text. When it comes to home security, organizing an active neighborhood watch and using multiple walkie-talkies for communication is a great way to handle perimeter security during a grid-down event, while using battery-powered alarms to take care of all interior security needs. Chapter 7 provides more detail regarding battery-powered walkie-talkies.

For those living in more rural areas, there are several low-cost driveway and remote storage building alarm systems with both the sensor unit and alarm receiver powered by either AA or C batteries. By utilizing rechargeable batteries as discussed in chapter 4, these alarm systems can operate many months without any grid power. Several different alarm types are available, including a remote sensor that can be buried under or beside a driveway, which activates only when a vehicle or large metal object passes by.

Other models are motion based, with the outdoor sensor located on a post or gate, and will alarm when vehicles, large animals, or people pass near the sensor. Slightly more expensive units allow using one battery-powered alarm receiver with multiple remote sensors and programming a different alarm sound for each remote sensor position, making it much easier to identify the location of the disturbance.

SECURITY LIGHTING

Security studies for remote commercial facilities have determined that a totally dark building or equipment yard that suddenly lights up with motion-controlled security lighting is a better deterrent than leaving all lights on throughout the night. This will be especially true during a grid-down event, when entire towns may be dark and all emergency power will be needed for interior illumination and not "wasted" on external lighting.

This means the sudden and unexpected activation of perimeter lighting around your home or equipment storage shed when an intruder approaches will be a significant deterrent during this increased level

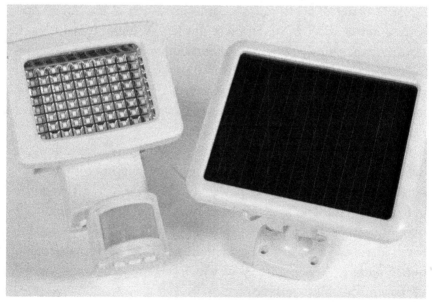

Solar-powered LED exterior floodlight with motion sensor

of vandalism and break-ins that typically occur on dark nights during extended power outages.

Most homeowners will already have one or more exterior floodlights around their garages and driveways that are motion activated or use a dusk-to-dawn photocell control. However, none of these will be functioning during an extended power outage. There are many different sizes of battery-powered outdoor LED floodlights that are motion activated and include a solar module to keep the battery fully charged.

Solar floodlight kits having a higher wattage solar module can power a light fixture containing 80 to over 100 bright LED lamps which will easily illuminate a large parking area.

Unit controls typically allow selecting from dusk-to-dawn, motion sensor only, or timed-off after a given number of hours to save battery charge. While you may already have multiple outdoor grid-powered floodlights, adding several motion-controlled LED floodlights that are totally solar-powered could be a real lifesaver when all other exterior lighting fixtures have gone dark.

I have not forgotten the early warning security a well-trained dog can provide. However, during an extended power outage it helps if you can safely see what's outside that's causing the early warning. Solar-powered LED floodlights will operate no matter how long the power is out, as they are always operating off grid.

SECURITY CAMERAS

Another useful battery-powered security device for a rural home or remote storage facility is a motion-controlled game camera. Normally sold for recording animals and game on a remote trail or around a feeding area, this is an ideal low-cost security system to record anyone trying to break into a storage building or steal your yard equipment while you are away. Since they are always battery powered, motion-controlled game cameras do not need the grid to work, and several models offer a solar charger option.

There are many sizes and features available, and most come with

Hidden game camera used for security

a tree-mounting bracket, which allows hiding the camera from others. Typically, you would locate the camera with the lens pointing toward the storage building or yard area you want to monitor. All game cameras have LED lamps that illuminate the area with invisible infrared light at night so the photo taken in total darkness will appear illuminated when reviewed later. Unfortunately, at night you can see these LED lamps, which glow a dull red color and will be noticed even if the camera is high up a tree. While the animals will not care, most accomplished thieves will be looking around for this red glow in the dark and will damage or steal your camera, along with whatever else they were after, to avoid getting caught. However, slightly more expensive game cameras use a type of LED lamps that do not glow and cannot be seen at night, which would be a better choice for your security-monitoring proposes.

All game cameras have weatherproof slots for the same memory cards typically sold for digital cameras. You will need an 8 MB size card or larger, depending on how often you check the camera, since most stop recording once the memory card is full. Most game cameras operate on four to eight AA batteries, which will record for several months, and several models use a separate 12-volt rechargeable battery pack for even longer stand-alone operation. Since game cameras are programmable, you can select from still photo or video mode, and how many seconds to allow between photos when activated by the built-in motion sensor.

While video mode or a higher rate of photos per minute does a great job of security documentation, this also will quickly fill up the memory card, especially if you have any deer or raccoons in the area that keep tripping the camera. Some trial-and-error testing may be required to find the correct balance between memory card size and frames per second.

Normally when I visit my remote storage facility, the first thing I do before unlocking the gate is to switch a blank memory card with the card in the camera, and replace the AA batteries. Hopefully I also remember to turn it back on again when I leave. I check the memory card later when I get home using my laptop and usually find a collection of deer and bear photos, but once I did catch the tax assessor snooping around!

EMERGENCY RADIO

An important part of security, especially during a grid-down event, is advance warning of any life-threatening events in your immediate area. During any type of major emergency, at least one AM radio station in each city is designated part of the National Emergency Broadcast System and will provide the same emergency information and evacuation advice to all affected areas. In addition, the National Weather Service (NWS) maintains a network of more than one thousand automated weather radio stations, covering all fifty states plus all coastal waters. These automated radio stations provide the first warnings of any severe storms approaching your area and recommended action to take.

The National Weather Service operates on seven specific frequencies in the 162.40 to 162.55 MHz high-frequency band, but these are outside the normal frequency range of any household AM or FM radio. You will need to purchase a specific NWS radio to receive these weather alerts, but these radios are small and very inexpensive. Battery-powered NWS radios are also easy to find. Since they have to be turned on at all times to hear the alerts, this is not a nuisance since just before any warning a special tone is sent, which activates the speaker in the receiver. Normally these radios make no sound at all.

NWS weather alerts can also be sent directly to any e-mail address and as a text message to any cell phone if you register on the NWS website. There is no cost for this service, and a test signal is sent every Wednesday afternoon at two specific times so everyone can verify that their radios are still operating properly. Having a battery-powered NWS radio can be a real lifesaver during any emergency, especially if you live in an area that experiences frequent tornadoes.

Normal weather updates for your area are broadcast on the hour and eleven minutes after each hour, but you can mute your NWS radio to stay quiet until it receives the activation code. When the activation code is received, your NWS radio's speaker is turned on and you will then hear an automated message describing the potential hazard. In addition to warnings for severe storms, hurricanes, and tornadoes, this same system is also used to notify residents about potential flooding, forest fires, dangerous ocean or river conditions, unusual solar activity that could disrupt communications, and even missing child (AMBER) alerts.

Each county of each state has an assigned code, and each type of alarm condition is also assigned a specific code. When you purchase a

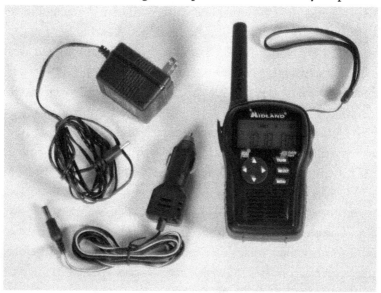

Battery-powered NWS portable weather radio and charger

NWS radio, you can indicate which counties are associated with your area so your NWS radio will activate only when one of these warnings is associated with your immediate area. In addition, you can also select which types of warnings you wish to receive.

While most NWS radios are tabletop design and include an alarm clock function as well as a basic AM/FM radio, for grid-down concerns I recommend one of the portable NWS radios. While these include a grid-connected power adaptor and charger base, you can take these radios with you if you do need to evacuate, and they are easy to use since they do not include the other non-NWS functions normally included with the tabletop models.

As soon as the radio is removed from the charger, it switches to internal battery power and will then start automatically scanning for the closest NWS radio station along your route. I really like the Midland HH54 and the Oregon Scientific WR602 portable weather radios, as both have long battery life when used out of the charger, and both use AA rechargeable batteries.

SMOKE DETECTORS

I am sure by now everyone has one or more battery-powered smoke detectors in their homes, apartments, and businesses. Not only do they really save lives, but they are very inexpensive and easy to install. While most will operate up to a year on the internal battery, it's good practice to change the battery when you change the clocks to daylight savings time, which serves as an easy reminder. The smoke detector is just another battery-powered device we depend on each day to protect our lives, but we still need fresh batteries when it's time to change, and these may not be available during a grid-down event.

In many larger homes and apartment buildings, the smoke detectors you see in the ceiling may not be battery powered. In fact, many are now powered by 120-volt AC grid power, and report an alarm condition back to a central alarm panel that is also powered by the grid. While these central fire alarm panels do have battery backup, the batteries are not sized to power these systems for an extended power outage, and without grid power the systems will soon stop providing fire alarm protection. If you find yourself living where this is the case, I suggest purchasing several low-cost smoke detectors that do not depend on grid power to operate.

BATTERY-POWERED CLOCKS
Most of us take for granted the telling of time. No doubt you have a digital wristwatch, bedside alarm clock, microwave oven clock, and video recorder clock. You also have a small clock display on your cell phone and laptop computer, and hourly radio time announcements. In fact, it's almost impossible to not know the time. However, all of these clock devices require the utility grid for power or to recharge. During a real grid-down event, most of these devices will fail or just stop working, and the simple act of knowing the time may not be so simple.

However, almost every office supply and furniture store sells large-dial digital wall clocks that are very inexpensive. Unlike all of the other clocks you own, these wall clocks are normally powered by one or two AA batteries and will run up to a year before needing a battery replacement. At least one large wall clock for the house that keeps perfect time without the utility grid or Internet is another must-have item.

While you are shopping for a battery-powered smoke detector and wall clock, this will be a good time to also purchase a quality A-B-C-type fire extinguisher for the kitchen, garage, and each vehicle, and be sure their locations are visible and easy to access. Working for years with our county's volunteer rescue squad, I have seen many small kitchen and vehicle engine fires that could have been easily extinguished with a small, handheld fire extinguisher if the owners had actually owned one. Instead, these small fires became big fires and soon engulfed their homes and vehicles.

During a grid-down event, your local fire department and rescue squad will no longer be just minutes away, and you could be on your own instead of getting the fast emergency help to which you are accustomed, so you'd better prepare now.

When reviewing the many ways to improve your security with battery power, I am hoping you realize that security will involve more than smoke detectors, motion alarms, and checking your door locks. Take time to review the types of hazards your home faces, and make sure you have a way to minimize each risk. This is also a good time to start a neighborhood watch program if you do not have one already.

This chapter has covered many different types of security concerns related to living through a grid-down period, and they are all important to consider. The battery-powered security devices I am recommending have followed the other chapters by standardizing on the AA and C rechargeable batteries. When it comes to security, remember that sometimes things just do not go the way you planned, so always have a backup.

15

USING VEHICLES FOR BATTERY POWER

Earlier chapters described how all types of portable battery-powered devices can be recharged from the 12-volt utility outlet in a car or truck. For this purpose I prefer a reliable older truck, especially the larger diesel models, as they typically have dual deep-cycle batteries. To increase mileage, most car designers need to reduce weight, and a heavy, deep-discharge battery and real spare tire are usually the first items to be jettisoned from today's fuel-efficient cars.

A deep-cycle truck battery has thicker lead plates, which can withstand a daily deep-cycle and can power a constant load for long hours. However, to reduce weight, most smaller car batteries have ultrathin lead plates and smaller overall dimensions since they typically only need to provide one large but short-duration discharge during starting, then a long period of recharging as the vehicle is being driven.

Charging any portable device using the vehicle's 12-volt utility outlet can quickly discharge a smaller car's battery unless the engine is running, and most vehicle alternators do not generate excess charging amps unless the engine is operating above its normal idle speed. While trucks typically have both a heavier battery and alternator, there is still the danger of discharging the battery below its ability to restart the engine when using the 12-volt utility outlet for any length of time while the engine is off.

There are several advantages with using a truck as a source of backup power during an extended power outage. Not only can multiple 12-volt DC devices be charged using the truck's utility outlet, but a small 120-volt AC inverter can also be plugged into this outlet to provide 120-volt AC power to those small electronic devices that cannot operate directly on 12 volts DC. I do not recommend using any inverter larger than 300 watts connected to a vehicle's 12-volt DC utility outlet, as above this power level the large amp draw can damage the outlet or blow the fuse.

An inverter in the 1200- to 1500-watt size range can provide emergency power for larger loads in the home, including a refrigerator or microwave oven using a long extension cord, but any inverter over 300-watt capacity needs to be connected directly to the battery terminals and not the dashboard's 12-volt utility outlet.

Obviously, fuel consumption and availability of fuel will most likely limit this temporary backup plan to only a few hours of daily use during an extended power outage. The cables from any inverter to the vehicle's battery should be less than five feet long and as large as car jumper cables. For a given load, the amp flow at 12-volt DC from the battery *into* the inverter will be ten times the amp flow at 120 volts AC *out of* the inverter.

For example, a refrigerator that draws 10 amps at 120 volts AC when it cycles, will cause the inverter to pull over 100 amps from the battery. It is also recommended to park the vehicle as close to the home as possible, and any extension cord from the inverter to the home should be at least No. 12 wire and labeled for exterior use. These heavier outdoor extension cords are typically yellow and will have a much lower voltage drop than indoor-type extension cords, especially in the needed longer length.

During a bugout situation, your truck can serve as a mobile emergency power supply, but your emergency preplanning should include replacing any vehicle battery over three years old using the largest group size battery that will physically fit. See the appendix for a table showing the dimensions of the most popular deep-cycle battery sizes. All batteries lose amp-hour capacity with age and lower temperatures, so if winter is approaching, even a three-year-old battery that works perfectly during the summer months will lose over half its capacity when the temperature approaches freezing.

The last thing you need is a vehicle that does not start during a crisis, so make sure the battery and charging system are checked regularly and the fuel tank is kept full just in case. In addition, parts suppliers offer alternators with a larger amp output to fit almost any model and year truck. Truck alternators and batteries are typically upgraded on vehicles ordered with power attachments such as a snow plow, emergency light bar, or winch, so if you have a reliable truck, even if an older model, making this electrical system modification and upsizing the fuel

Deep-cycle RV battery and 600-watt inverter

USING VEHICLES FOR BATTERY POWER

tank will turn your truck into a great "get out of Dodge" vehicle, especially if it has four-wheel drive.

Some preppers mount an inverter inside the truck cab and use permanent heavy-duty battery terminals and cables to make the electrical connection to the battery. I have a 1500-watt inverter mounted behind my truck seat, along with a separate heavy-duty extension cord, electric chain saw, and portable light. If I need to remove a tree limb blocking the road after a storm, I never need to worry about a gasoline chain saw that leaks oil or will not start, or a battery-powered chain saw with a dead battery.

If you temporarily end up living out of your truck during an emergency evacuation and need to charge multiple electric devices each day, I recommend taking along a higher-wattage fold-up solar module to keep your vehicle battery charged when parked. These include grommets in each corner. Four bungee cords will easily lash the solar module across the engine hood or roof of cab. Orient the vehicle so the solar module faces the sun and is not shaded by nearby trees or buildings. However, a flat mounting is acceptable during summer months when the sun's path is more directly overhead.

A heavy-duty four-wheel-drive truck is a must-have for the serious prepper, and older models with a few dents will be less conspicuous and less likely to be stolen, while hiding a very reliable power train, extended-range fuel tank, and robust backup power supply.

16

RV CAMPING WITH BATTERY POWER

RECENT WINTER STORMS HAVE HIT OUR area hard, causing several power outages lasting days. An acquaintance and I were discussing the effects of one of these power outages. My friend told me about not having any heat in their all-electric home, no way to cook meals, and no way to wash up, as they were on a well system. Since we both have recreational vehicles, I asked why they didn't temporarily move into their RV parked in the driveway for a few days, and they both looked at me like a deer caught in the headlights. The thought had not occurred to them! The space heater, kitchen stove, refrigerator, backup generator, and hot-water tank installed in all RVs operate on propane, and most RVs and truck campers include two propane tanks with automatic switchover when one tank gets empty.

All RVs also have a dual electric system, with all light fixtures, flat-screen television, DVD player, radio, water pump, and the fan for the gas furnace all powered by the 12-volt DC batteries. Some larger RVs also have a satellite dish for cable television and high-speed Internet connection, which are also 12-volt DC powered. The motorized leveling jacks on any motor home or truck camper are also powered by 12-volt DC motors.

While most pull-behind RV trailers and fifth wheels require plugging into grid power at an RV park to power the 120-volt AC air conditioner and microwave oven, most motor homes and truck campers have on-board generators, which can power the air conditioner and microwave if you choose to dry camp without staying at an RV park. Since these appliances rarely need to operate for long, the generator is usually turned off most of the time, with everything else still working on battery power.

While I have owned several different RVs over the past few years, my favorite by far is our current slide-out truck camper. Truck campers with a slide-out are much larger and heavier than campers without a slide-out, and normally require a long-bed pickup truck with dual rear wheels. My F-350 diesel crew cab not only easily handles our truck camper, but also provides lots of additional space for extra supplies and camping gear.

I originally thought our thirty-two-foot-long RV and pulled-behind trailer would be the perfect way to travel if we ever did have to evacuate. However, after several years of skipping from one RV park to the next for the needed utility hookups each night and requiring a football-size parking lot to maneuver up to the gas pumps, I finally realized a truck camper on a four-wheel-drive pickup truck was the best bugout vehicle available. Although my rig slightly overhangs one normal parking space when parked, it is still much easier to maneuver than an RV trailer or fifth wheel.

An on-board generator means any paved parking lot or off-road gravel area along a rural back road can be our RV park since we require no hookups. All RVs have three water tanks under the floor: one for fresh water, one for gray water (sinks and shower drains), and a black water tank (toilet only). While the size of each tank varies with the size of the RV, our camper's fifty-gallon tanks mean we can go several days without connecting to water and sewer hookups supplied by an RV park.

To make any RV more suitable for emergency evacuation, there are several modifications you can do yourself or have completed by a local RV service center. First and easiest is to retrofit all light fixtures with new LED bulbs or new LED light fixtures. Most RVs are shipped with 12-volt DC incandescent lamps in all light fixtures due to their lower costs, but they are a major drain on the batteries, especially when dry camping. If your RV has a typical 12-volt RV/marine battery, I would replace it with two heavy-duty 6-volt deep-cycle batteries.

Being able to travel and camp several nights without having to plug into an RV park requires a higher amp-hour battery capacity. Most RV battery compartments have the space for dual batteries, although the second battery is rarely included to help lower the initial sale price.

I also suggest installing a solar module or multiple solar modules on the roof of your truck camper or RV. Since the roofs of most RVs are covered with skylights, a satellite dish, radio antenna, air conditioner, and exhaust fans, a higher-capacity solar system may require multiple smaller modules to miss these roof obstructions. While I have used motorized tilt mounts so the modules can be raised to face the sun, I found most of the time the RV ended up facing the wrong direction to use a tilt mount.

Truck cap with 100-watt solar module

Therefore, mounting the solar module(s) flat, with a one-inch air space between the modules and the camper roof, provides almost the same performance regardless of which way your truck or trailer faces.

Many new models of RVs are shipped "solar ready" with a weatherproof connector on the roof and spare wiring routed down behind an empty wall space for a future solar charger controller, then tied off in the battery compartment. To add solar you need only to mount the

Truck camper with dual 125-watt solar modules

solar module(s) on stand-off roof brackets, connect to the weatherproof plug, install the charge controller in the wall where the solar wires pass behind, and connect to the batteries. Having 200 to 300 watts of solar modules on the roof of a truck camper or RV trailer can provide all your power requirements indefinitely, except for the air conditioner and microwave oven, which still require running the built-in generator or using an RV park hookup for grid power.

My only caution is every solar-ready RV that I have modified included factory-installed "solar" wiring and a roof plug rated for only 15 amps of solar capacity. At 12-volt DC, this limits you to a solar array in the 200-watt range, so installing a much larger solar array for serious dry camping will require replacing this smaller "solar ready" wiring with No. 10 size or larger wire, and replacing the roof plug with a weatherproof junction box.

If your budget is closer to a pull-behind fold-up camping trailer or truck cap instead of a full-size RV, there are still solar conversions you can make. Two solar modules in the 100-watt range each can be hinged together along one side if they include a sturdy aluminum frame. This allows double the wattage in the same storage space, and these can be pulled out of a storage compartment, unfolded, and tilted against the camper or truck facing the sun, which will be south or southwest during the early afternoon. Ground stakes or bungee cords can be used to protect the solar modules from high winds.

All that is left to do is unroll the attached cable and plug into the batteries, and there are many types of DC quick discounts available for this purpose. You will need a solar charge controller, but this can be permanently mounted on the back of a solar module or in the battery compartment. These solar components are discussed in far more detail in chapter 5. Being able to travel or stay in an RV can be a real lifesaver after a disaster, or just camping in your driveway during the next power outage.

17

BUGGING OUT WITH BATTERY POWER

OKAY, NOW IT'S TIME. YOU FOUGHT the good fight and have stayed home as long as you could using a generator or backup solar system, but now you have to leave. While it's usually better to stay in your own home during any emergency as long as you have made the proper preparations, sometimes people are forced to leave their homes during a disaster and may never be able to return. Watch the evening news; this is happening somewhere in the United States every single day due to flooding, forest fires, mudslides, tornados, or chemical spills.

This could also mean you will be on foot with only what you can carry in a backpack. While traveling in a well-stocked RV or truck camper would be ideal, as discussed in chapter 16, this may not be possible when a real crisis hits. For example, you could be hundreds of miles away on a business trip, at school, or working at a distant factory. When a grid-down event occurs, it may be far safer to head for home or an alternative safe meeting place immediately, before the traffic gridlock or gang violence begins. It's also possible all public transportation has already stopped and driving is no longer a reality.

BUILDING A BUGOUT BAG

"Bugging out" means fleeing in an emergency. It is not the same as going camping for several days. A bugout bag is a prepared bag that includes only emergency supplies. It can be a book bag or a gym bag and you should include items such as:

PERSONAL WATER FILTER	ENERGY BARS
BATTERY-POWERED AM RADIO WITH EARBUDS	FOLDING KNIFE
NYLON ROPE	BATTERY-POWERED LED HEADLAMP
BATTERY-POWERED LED FLASHLIGHT	THERMAL BLANKET
CELL PHONE CHARGER AND ADAPTER	MAPS OF CITY AND HIGHWAYS
WATERPROOF MATCHES AND LIGHTER	CANNED HEAT (STERNO) OR CANDLES
FOLD UP RAIN SLICKER WITH HOOD	WHISTLE
SIGNALING MIRROR	COMPASS
TOILET PAPER	HAND SANITIZER
TOOTHBRUSH AND TOOTHPASTE	DUST MASKS
BASIC FIRST-AID KIT	BENADRYL
DUCT TAPE	

A book bag or gym bag is a great way to carry the few survival items you will need. This is especially true if you have to bug out from an office or apartment in the city, since nobody would expect survival supplies or anything of real value to be carried around in a book bag. However, a loaded-down military-style backpack will not stay on your back long when in a panicked crowd running down a city street, trying to escape some disaster.

BUGGING OUT WITH BATTERY POWER

There are endless articles concerning what to pack in a bugout bag, but I like to keep things simple. I would much rather have identical backpacks containing the same basic supplies stored in every vehicle I own, plus a spare backpack at work, than to own the perfect backpack holding everything but sitting fifty miles away from where I am when I really need it. Keep in mind that backpacks for emergency evacuation are only to help you get from where you are now to where you need to be, so keep the size and weight to a minimum.

Each backpack I have contains some bottled water, a water purifier bottle, several high-energy snack bars, two dust masks, a highly compacted rain slicker with hood, a thermal space blanket, a quality LED headlight with head strip and extra batteries, one medium-size folding knife, a roll of parachute cord, a fold-up street map, a Bic lighter, a stainless steel coffee cup with a long handle, canned heat (easily fits inside a coffee cup), fire-starter gel packs, a referee whistle, a few bandages in various sizes, and a small mirror.

Space and weight will be at a premium, but you should still have room for a battery-powered all-band radio and earphone, a cell phone charger, and a portable GPS device, and these need power to recharge after use. You should have a 12-volt DC adaptor for each of these electronic devices, which allows recharging from the 12-volt DC utility outlet in the vehicle. However, if you are on foot or your vehicle is out of fuel, you will still need these electronic devices to function or you will soon be lost, in the dark, and not know what is going on beyond your immediate location.

You will want to pack at least two fold-up solar chargers with different wattage

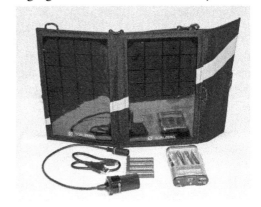

Fold-up Goal Zero 7-watt solar charger comes with various adapters.

ratings. Smaller fold-up 5-watt solar chargers are perfect to keep your cell phone, GPS receiver, or LED flashlight charged. A larger foldout solar module in the 25- to 40-watt range can quickly recharge tablet computers and two-way shortwave radios if you have the extra space.

Any electronic device having an internal rechargeable battery that includes its own 12-volt DC adaptor can easily connect to most foldout solar chargers or your vehicle's 12-volt utility outlet to recharge.

Along with each of my basic bugout bags I keep a pair of really good hiking boots and two pairs of hiking socks. These are not tennis shoes or sweat socks; they were selected for having the right design for miles of walking, which may be the only form of transportation available when a grid-down event first hits. Each pair of hiking shoes received multiple applications of a quality boot waterproofing spray as soon as I got them home from the store. I also wear each pair several days to break them in before packing them away with each bugout bag.

The last thing you need during an emergency is to hike ten miles in a brand-new pair of shoes or boots. I am sure others will suggest a different combination of must-have emergency supplies in their backpacks, but remember: this bag and these hiking shoes are just to get you to a much safer area and are not intended to supply a normal camping trip. A true evacuation may require walking many miles, and it could be cold, raining, and dark.

Don't assume an evacuation will involve a sunny spring day on flat ground. Many preassembled survival kits are filled with inexpensive gadgets just to look impressive, but these may have little or no use in a real emergency. I suggest making up your own bugout bags, using this book as a guide.

18

CONNECTING DEVICES TO BATTERY POWER

WHEN YOU WANT TO POWER YOUR conventional television, kitchen appliance, or hair dryer, you just plug it into the nearest 120-volt AC wall outlet, which by code are spaced no more than twelve feet apart along every wall. You don't need to think about how the electricity got there, where it was generated, or if there will be enough additional capacity to power your device. It just works—at least until it doesn't.

Most European countries have standardized on a nominal 220- to 240-volt fifty-cycle AC electrical power using a two-prong, ungrounded wall receptacle,

Don't forget the adapters!

147

although the actual size and spacing of the two-prong plugs varies from country to country. However, the United States standardized on a nominal 110- to 120-volt sixty-cycle AC electrical distribution from the very beginning, although after the 1950s we added a ground pin to all receptacles and larger-wattage appliance plugs. Unlike Europe, all of the wall outlets in the United States use a standard spacing of the outlet and plug blades, although one blade is slightly wider, ensuring the plug will be oriented correctly so the neutral blade is connected to the neutral conductor.

Most small household appliances have housings made from nonconductive plastic that provides a double layer of electrical insulation, which allows using a two-bladed polarized plug. However, most of the larger appliances with metal enclosures now use the three-bladed plug having an added ground pin to bond all non-energized metal parts together to reduce the risk of shock. All two-blade and two-blade-with-ground-pin appliance cords will fit into the same 120-volt AC wall outlets.

Battery-powered devices are different from grid-powered devices, as most operate from internal batteries and are plugged into a power-source only to recharge. However, battery-powered devices will operate on external power when their chargers are left connected to a grid-powered outlet, which saves their internal battery charge. To reduce the size of a portable device and accommodate the different grid standards, most battery-powered electronic devices use a combination power cord and in-line battery charger or plug-in "wall wart."

This can be a nuisance since you now need to have a separate charger and dangling power cord nearby for each battery-powered device whenever it needs to be recharged. However, having a separate, external charger for each device does allow using multiple sources of power to recharge the same device, as in-line chargers are available to match almost any power source.

Unlike an AC appliance, which will only operate when connected to a 120-volt AC wall outlet, any battery-powered device can be recharged from any 12-volt DC utility outlet in a car, truck, boat, RV, or solar

CONNECTING DEVICES TO BATTERY POWER

module, as well as a 120-volt AC grid or generator power when available, as long as you have the matching charger. The downside is you will need a different charging device for each power source, and there are different sizes of plugs used to connect the charger to each battery-powered device, making it harder to use the same charger for multiple devices.

During a true grid-down event, flexibility is very important, and you will likely be called upon to do lots of improvising to keep all your battery-powered devices operational. To make this easier, it's important to know the different types of plugs, connectors, cables, and adaptors currently used by all battery-powered devices. It's also helpful to search for those electronic devices that use the same connectors, which can share the same battery chargers. Also remember that wire size is determined by the amp flow it must handle, not the voltage, and for a given watt load, it takes ten times more amps at 12 volts than at 120 volts, so DC wiring will be a larger gauge to supply the same wattage load.

Let's first review the types of battery connectors commonly used. Most of the rechargeable LED lanterns, portable radios, and smaller electronic devices that use rechargeable batteries have a "barrel" connector to connect an external charger.

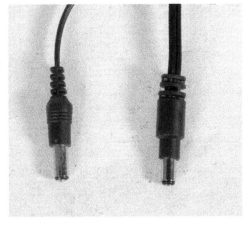

Barrel-type DC connector plugs

Unfortunately, these are manufactured in multiple diameters, different lengths, and even different sizes of the internal pin. While most connectors are polarized, with the center pin positive and the outer shell negative, you should verify this before using homemade adaptors, as this standard is not absolute. For many smaller rechargeable 12-volt DC devices, the barrel will be either 2.1 mm diameter by 5.5 mm long, or the larger, 2.5 mm diameter by 5.5 mm long. Typically the other end of the adaptor cable will be either a wall wart for a 120-volt power supply, or a cigar plug, also referred to as a cigarette lighter plug, to fit a 12-volt DC vehicle outlet.

The 12-volt DC charging plug used for almost all battery-powered devices will be a cigar plug to fit the vehicle utility outlet. Since just about every car built since the Model T included a dashboard cigarette lighter, today's vehicle manufacturers still include the 12-volt DC socket to power all of the many 12-volt DC accessories available, but they no longer include the actual cigar or cigarette lighter. No doubt most of you have a 12-volt DC cell phone charger or GPS receiver that you have plugged into your car's dashboard utility outlet right now.

My truck came with three different 12-volt utility outlets, and my car has a 12-volt utility outlet in both the front and back seats, plus an

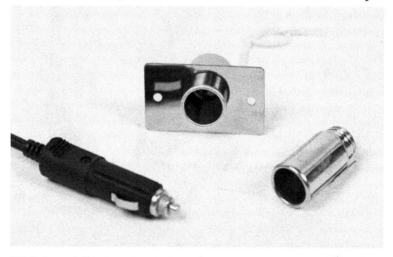

Vehicle "cigar plug" and 12-volt DC utility outlet

additional one inside the center console. Boats, recreational vehicles, ATVs, and truck campers are also now furnished with multiple 12-volt DC utility outlets.

Since you will have multiple battery-powered devices that will need to be charged at the same time during a real emergency, I recommend keeping several multi-outlet 12-volt DC adaptors in each vehicle and RV. Most cell phones, rechargeable flashlights, and portable radios rarely draw more than one or two amps at 12 volts DC when charging; therefore, charging multiple devices from the same 12-volt DC vehicle outlet should not be a problem, as most vehicle outlets are typically rated and fused for 15 amps.

Having multi-outlet adaptors will also save generator fuel or solar charging hours as opposed to charging everything one at a time, which would be required without a multi-outlet adaptor. Remember: your home has 120-volt AC outlets on every wall of every room, but you may be limited to a single source of 12-volt DC charging power during an emergency. You should not modify 120-volt AC extension cords to connect 12-volt DC devices. Use extension cords specifically sized and rated for 12-volt DC applications, and preferably with an outdoor-rated insulation.

SAE style battery connector

Some manufacturers of fold-up solar modules have standardized on the "SAE" quick-disconnect 12-volt DC plug and receptacle to connect their solar module to other devices. These are fairly easy to recognize, as they are flat molded plastic with the positive and negative pins side by side, but one pin is exposed. To avoid a mismatch of battery polarity, they cannot be plugged together with the pins reversed. However, it can be difficult to identify which pin is positive and which is negative if you are making up your own adaptors, so verify this with the nameplate on the device.

With the tremendous increase in the use of portable laptop computers,

5-Volt DC "USB" power and data connector

iPads, and smartphones, most manufacturers now furnish a USB-type connector on the end of the cable between the battery-powered device and the source of charging power. Manufacturers usually provide both a 120-volt AC power supply and a 12-volt DC adaptor, with each sharing a common USB-type plug. This allows using the same charging cable regardless of power source. Note that all USB outlets are 5-volt DC, not 12-volt DC. Any adaptor you use to connect a USB cable to a 12-volt DC socket must include a DC-to-DC voltage converter built into the plug or in-line charger to match the different DC battery voltages.

Most 12-volt DC utility outlets in vehicles are sized to charge devices with only a few amps of current draw at 12 volts DC. However, there are some 12-volt DC-powered devices that require far more power to operate and do not have an internal battery, so they draw full power from the vehicle battery at all times they are operating. These can easily exceed the 15-amp rating of most cigar-style 12-volt DC vehicle outlets. These larger loads should be wired directly to the vehicle's battery using an in-line high-capacity fuse and ring-type battery post terminals.

It is not uncommon for portable CB and shortwave radio transmitters to draw up to 30 amps of 12-volt DC power, and most home radio base stations are actually mobile radio units connected to a high-capacity 12-volt DC power supply operating from a 120-volt AC wall outlet. The amateur radio community has a long history of helping out with emergency communications during national disasters, with lots of experience operating from emergency generators and backup battery systems.

As mentioned earlier, there are several national volunteer organizations offering training and official recognition for ham operators, including the American Radio Emergency Service (ARES) and the Radio Amateur Civil Emergency Service (RACES) networks.

To address the problems of interconnecting their many different radio and transmitter combinations to multiple sources of emergency backup power, these amateur radio groups have standardized on the two-conductor red and black genderless in-line connectors commonly referred to by the trademark name Powerpole, which are available in 15-, 30-, 45- and 75-amp capacities using polarized 12-volt DC red (positive) and black (negative) conductors. When viewed from the mating end with the "hood" protector oriented on top, the positive (red) conductor is connected to the left side, and the negative (black) conductor is connected to the right side.

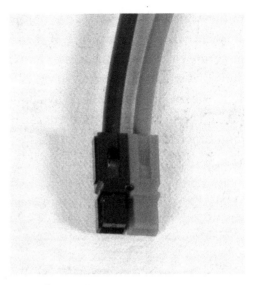

Powerpole-type DC connector

Since a special crimper is required to connect the wire to the plug's internal connector blades, it is best to purchase these connectors premade. However, they are fairly inexpensive if you want to make your own, with the 15-amp Powerpole wired with No. 16 to No. 20 wire, the 30-amp wired with No. 12 to No. 14 wire, the 45-amp wired with a No. 10 to No. 12 wire, and the 75-amp Powerpole wired with a No. 6 or No. 8 wire.

Most manufacturers of portable shortwave radios require using a separate in-line fuse in both the positive and negative connections to the vehicle's battery for added protection. Regardless if you are an amateur radio operator or not, if your emergency preparations include powering electrical appliances or tools having a higher 15- to 75-amp draw that would overload a typical 12-volt DC vehicle outlet, you can avoid "reinventing the wheel" by taking advantage of these ARES and RACES connector plug standards established for all DC-powered communications equipment.

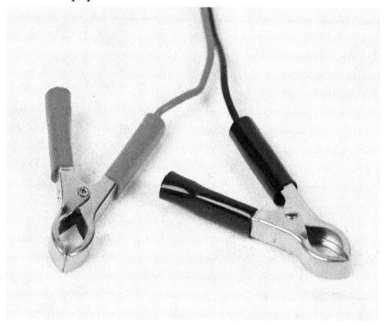

"Alligator" battery post clamps

To temporarily connect a portable DC to AC inverter to a 12-volt DC vehicle battery, most inverters over 300 watts are furnished with "alligator"-type spring connectors for direct attachment to the posts on the vehicle battery. However, this should be considered a very temporary type of battery connection and is not intended for continuous operation.

For example, most vehicle batteries are either difficult to access in smaller car models, or may have terminal corrosion, which can make it difficult to achieve a good electrical connection. In addition, this type of battery connection will not allow closing the hood, and the battery cables could tangle with the engine's cooling fan, as the engine will most likely be running when a larger-capacity inverter is being powered.

If the battery post connection does not make good contact, the alligator clip can become very hot. My advice is to use a ring-type cable end to attach directly to the battery posts if the inverter exceeds the 15-amp load limit of the 12-volt DC dashboard utility outlet.

While wire colors may mean very little to you unless you are an electrician, it is important to follow recognized wire color codes. This is especially helpful and safer if others may need to help install or are called upon to repair any electrical power wiring. While the *National Electrical Code* specifies using black for all ungrounded power conductors, white for neutral conductors, and green or bare wire conductors for grounding, most vehicle, RV, and boat wiring still use the older standard of red for positive and black for negative.

For practice I suggest placing all of the battery-powered devices you own on the floor and then trying to map out how you will keep each charged during a real grid-down event. Most likely, the first few weeks may allow using a portable generator as your power source as long as you have a reliable fuel supply, so you will need connectors and chargers for 120-volt AC. Limited fuel may require switching to a vehicle utility outlet or a solar charger, so you will also need cables and chargers for 12-volt DC.

This is another reason to have the right adaptors to allow charging multiple devices from the same charger. Consider multiple charging arrangements, and assemble the DC adaptors, plugs, and low-voltage cabling you will need to make this work later under potentially very stressful conditions. Visualize which devices may need to go with you if you have to evacuate, and make sure to have the required adaptors and chargers available in each vehicle.

While I may have a few more battery-powered devices and associated chargers than most due to my product testing, I was having a very hard time matching up the correct charger, plug, and charger cable to the correct device. This was especially true when packing away multiple devices together in EMP-protection storage cans. They all are similar in size and shape, and are usually solid black. It is also hard sometimes to identify their voltage and charging capacity markings and, while their connecting plug may fit the device, it may have the wrong plug polarity or charging voltage.

I use a silver-color magic marker to write on each charger the name of the device it serves. It's surprising how much time this saves, and it keeps you from arriving at your destination with a mismatched plug, cable, or charger for the battery-powered device you had intended to use.

As you acquire more battery-powered devices in preparation for a major power outage, you also will end up with a large assortment of different types of charging adapters, having different types of cables and plug ends, to provide different device charging voltage requirements. While things are still fairly normal, organizing all your chargers and cables now will provide a much less stressful time during the next extended power outage. A little pre-planning, organizing, and trial runs will go a long way toward reducing stress levels when everyone else is in panic mode.

19

COOKING WITHOUT GRID POWER

IF YOU ARE TRYING TO STAY in a conventional home during a grid-down event, you will not have utility power to operate any electrical cooking appliances, and any backup generator you expect to keep appliances working will soon run out of fuel. This means no electric water heater, no electric stove, no electric toaster, no electric coffeepot, no electric oven, no electric griddle, and no electric microwave oven.

If you have gas cooking appliances supplied from a local gas utility, you may or may not have gas for long, depending on the ability of the utility to maintain acceptable line pressure throughout their distribution system without the grid to power large gas compressors, distribution valves, and their computerized controls. Having a barbecue grill on the porch and several portable tanks of propane is a very useful backup plan for meal preparation, at least for as long as the propane supply lasts.

Most off-grid solar homes have a propane cookstove and propane or solar water heater to reduce the size and cost of their solar-power array and batteries. However, once the propane, kerosene, or heating oil tank is empty and may not be refilled for months, what are you expecting to use to cook your meals and heat the water? Any type of electrically powered heating appliance requires far more power than you can supply from even a roof covered with solar modules, so you will need to find another solution for your cooling and heating needs during an extended power outage.

We enjoy the convenience of a large commercial propane cookstove in our own kitchen. We also have a wood cookstove next to the kitchen stove that can cook food and boil water, as well as provide some space heating while cooking a meal.

Although a microwave oven is a large electrical load on any backup power system, the short cooking time of minutes, not hours, is an acceptable drain on most portable generators during an emergency, at least for as long as you have fuel. However, all but the smallest microwave oven will need a generator or inverter with at least a 1,500-watt output.

Avoid using any microwave that includes digital program functions and a lighted display during a grid-down event, as it will be a constant drain of electricity twenty-four hours per day. It's still possible to find a small microwave oven with a mechanical-dial timer and no electronic digital controls. These are typically sold for commercial food-reheating applications and do not consume any power after the timer shuts off. In addition, these are less susceptible to EMP damage since there are no digital displays or microelectronic components.

If your kitchen microwave is a large-capacity model with a digital display and multiprogram functions, keep a small backup microwave oven in reserve for those times you are on generator or backup power. These are fairly inexpensive and will easily fit in any closet until needed.

You may want to stay in a conventional home during a grid-down event, but you still need to eat, and this necessitates a way to cook meals that does not require the local gas or electric utility. Even if you are down to only freeze-dried foods and dehydrated soups during an extended power outage, you still need to boil water to prepare these simple meals.

In the end, though, you may have to lock up and bug out, but you will still need a way to cook your food. One of the most practical portable cookstoves I have found is the Volcano Grill/Stove.

The Volcano Grill/Stove burns charcoal, wood, and propane.

While it is designed to be fueled by a small amount of charcoal, it will easily burn wood and also includes a removable propane burner if you still have propane. Unlike most portable camping stoves, this stove can support and heat normal-size kitchen pots and heavy cast-iron skillets, while easily folding up to the size of an overnight bag. A dozen large bags of unopened barbecue charcoal can be stored indefinitely if kept dry, and will provide up to six months of daily meal preparation and boiling water using this stove.

If you must stay in a small apartment during a grid-down event, you can still store several weeks' worth of dehydrated meals, soups, and drinking water under a bed or in a clothes closet.

However, you still need to boil water to prepare most meals, yet you will most likely not be able to use a charcoal or propane barbecue grill in an apartment or multiunit condo. I recommend buying one of the single-burner butane stoves, which are safe to use indoors. You typically see these in restaurants for tableside cooking applications or at the end of a cafeteria-style food line for frying special-order eggs.

These butane stoves are very inexpensive and are designed to be easily folded up and packed away in their lightweight case, which will easily fit under a bed. Be sure to have plenty of extra fuel canisters, as these are inexpensive and do not expire. These stoves would also be handy in your bugout vehicle, along with a pack of disposable butane canisters.

Camping supply stores offer all types of small, fold-up cookstoves that are lightweight and will easily fit in a bugout bag if you do have to evacuate. The least expensive way to heat water is by a stainless steel cup with an extra-long fold-up handle, held over a can of Sterno fuel. These are typically sold for food warming trays, and the fuel can fits just inside the cup, along with a butane lighter or stormproof matches, for easy storage. Being able to boil water makes it safe to drink and easy to prepare freeze-dried meals.

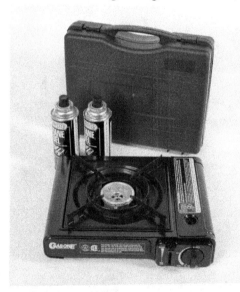

Portable butane stove with case

20

WASHING WITHOUT GRID POWER

MOST OF THIS TEXT HAS ADDRESSED multiple ways to use battery power to provide the same operations that are normally powered by the grid. However, there are several everyday activities, including cooking, discussed in the previous chapter, that do not require either battery or grid power to operate.

Washing clothes today is an almost autopilot type of operation that gets repeated every few days, month after month, year after year. Sort into piles, toss into washer, add soap, press start. Thirty minutes later, pull the damp clothes out of washer, toss into the dryer sitting two feet away, and press start. Come back when the buzzer sounds, take out and fold, and you're done. Easy, at least until the grid goes down.

While almost everyone can get by for a week or two without a washer and dryer by digging a little deeper in their less-favorite sections of their closets and wearing things several days in a row, at some point something has to give. Until now most power outages typically end just about the time you run out of clean clothes, towels, and sheets, but what happens when the power is still out? How do you wash and dry clothes without running water and electricity?

It is estimated that 82 percent of all American homes have an automatic clothes washer and wash more than three hundred loads per year. Most washers require between twenty-five and thirty gallons of water per cycle, which represents over 20 percent of a home's entire annual water consumption. While many apartments and condos do not have space for their own washer and dryer, rows of coin-operated washers and dryers are usually waiting at the bottom of the stairs or down the hall. Of course, there is always a commercial laundry available in every town.

Early turn-of-the-twentieth-century clothes washing was basically a hand-powered process, consisting of a wooden washtub and a paddle or washboard. A woodstove heated the water in an oversized kettle, although some later models with metal tubs were mounted over a firebox. Eventually, hand-cranked devices or a manually tilted tub were added to move the clothing back and forth through the soapy water.

Draining and refilling with fresh water provided the rinse cycle; then twisting or rolling up the clothes by hand removed most of the water before heading out to the clothesline. This water-removal process was eventually replaced by hand-cranked rollers, which would squeeze the water out of the clothes. During the 1940s the wringer washer was invented. It had an electric motor to power both the tub's agitator and the wringer rollers.

I am not saying that when the grid goes down you should start banging your wet clothes on a rock while standing in a nearby stream. However, there are new products on the market that take the early hand-operated methods of clothes washing into the twenty-first century using modern materials. These products are easy to find with an Internet search, and the costs are very reasonable. Having a backup plan to replace your automatic washer and dryer when the power is out for weeks or months will make life a little more bearable, or at least your family will smell good!

A commercial twenty-six to thirty-five-quart mop bucket with a side-mounted wringer can serve as a hand-powered clothes washer. (Please do not actually use this mop bucket clothes washer as a mop bucket!) A manual clothes agitator, which looks like an oversized commode plunger, is used to agitate the clothes placed in the (hopefully) clean bucket full of soapy water. The plunger-looking device is designed with multiple openings and layers to match the function of the agitator in a regular washing machine. While this hand-powered clothes washing operation is harder than pressing the start button on a powered washing machine, the clothes will actually still end up almost as clean.

You can use a second bucket for the rinse cycle if you do not want to switch from soapy to rinse water in a single bucket, while using the mop wringer to squeeze the water out of the clothes. These janitorial mop buckets are larger and more durable than a typical home mop bucket, and their attached wheels make them easy to roll out of the way when not needed. I have found these commercial mop buckets with attached wringers in almost every builder and janitorial supply outlet.

The next step up from the mop bucket washing machine is based on using a galvanized oval washtub, usually found in a farm or garden supply store. With a little more searching, you can also find these oval-shaped galvanized tubs with a center divider, allowing soapy water to be in one side and rinse water in the other.

A separately purchased hand-cranked clothes wringer is attached to the side of the tub you are using. Yes, believe it or not, you can still purchase a new hand-crank clothes wringer that has clamps to attach to the top rim of almost any tub or utility room sink. My Internet search found several quality hand-cranked wringers for slightly over one hundred dollars. While these may look like something your great-grandmother used a century ago, these are actually new designs using modern, rustproof materials. The wringers have an adjustable bottom diverter tray allowing the water to be squeezed out of the clothes, then returned to the rinse water or discarded over the side.

While ironing will be the least of anyone's concerns during a

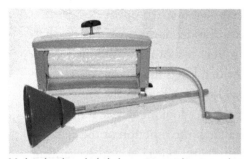

Modern hand-cranked clothes wringer with agitator that you can use with any washtub or utility sink

weeklong power outage, being without power for a month or more is a totally different proposition. It's still fairly easy to find several models of "sad irons," called such because *sad* is Old English for "heavy." These are sold in pairs: one is being warmed on top of your woodstove while the other is being used; they are switched when the one you are using starts to cool. Although the heavy metal base can get really hot if needed for cotton shirts and jeans, the insulated handles slide on and off and are not left attached while heating up. Using these sad irons will give your arms a good workout, but they will do the job and require no electrical power.

Sad irons, like the washtub and clothes wringer I described for cleaning clothes without electricity, will require buying now if you see this as a need for your family. Trying to purchase any emergency preparedness appliances, foods, and solar chargers after a grid-down event occurs will be like trying to buy bread, milk, and batteries at your local grocery store the day before a hurricane. Not available at any price.

When it comes to clean-smelling bedsheets, towels, and clothes, all the washer and dryer scent additives in the world will not provide the fresh smell of clothes that were dried on an outside clothesline. In addition, the ultraviolet spectrum of sunlight combined with a small breeze is a natural disinfectant.

Ever wonder where all that lint comes from that

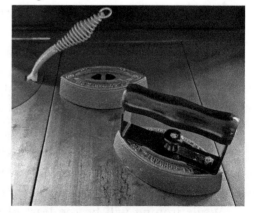

Clothes ironing without electricity

you have to remove after each dryer cycle? Drying clothes on a clothesline puts less stress on the material, buttons, and seams than hot-air clothes dryers. It will not slowly convert your clothing into lint or weaken the threads used to sew the seams.

My own family and many of our friends own perfectly good gas or electric clothes dryers that we have stopped using except during the coldest weeks of the year. Instead, we have installed clotheslines just outside our utility rooms in a nice, sunny area of our yards. Packaged clothesline kits are available from any builder's supply outlet. They include two metal poles, a roll of clothesline, and turnbuckles to keep the lines tight.

The kits are easy to install, as you only need to dig two holes, pour in a bag

Solar-powered clothes dryer

of dry concrete mix, and add water. You can also make your own posts and crossarms from any Schedule 40 steel pipe, with two-inch minimum pipe size. I tried using a smaller pipe size but ended up having to install "guy-wires" at each end to prevent the poles from bending.

The spacing of the poles should be no more than twenty feet apart. (A wider distance can result in too much center sag with heavy bedsheets.) With a typical five-line configuration, this provides up to one hundred feet of clothesline.

The next time you take a drive through the countryside, look around; you will be surprised how many "solar" clothes dryers you see in all those backyards, and every single one will work without the electric grid! This is another useful project you can install now that will lower your utility bills every month, while silently standing ready for a major grid-down event.

PORTABLE SHOWER

My favorite new battery-powered product for grid-down bathing is the Pure-Clean battery-powered bucket shower. This consists of a waterproof pump and battery unit about the size and shape of a large pear, connected to a showerhead with a six-foot-long hose. The pump unit includes a waterproof switch and charging port, and is dropped into a five-gallon bucket of warm water as soon as the start button is pressed.

The showerhead is almost identical to the low-flow showerheads in RVs and truck campers, and provides plenty of uniform water spray. While you won't have enough shower time to go through your favorite show tunes, five gallons of water will easily last six minutes, which is more than enough time for a quick body and hair wash. The unit will provide multiple showers before needing to be recharged, and both a USB charging cable and a separate 120 VAC charger are included.

The water can be supplied from any nearby spring, stream, or lake and heated on a wood stove or backyard grill during a power outage. Although some may try showering in the driveway with a bag over their head, you could easily place the bucket and pump unit in your home's shower stall or bathtub when a power outage has shut down your well pump or city water supply. Being able to take an occasional shower during a power outage lasting weeks can provide some creature comfort during an otherwise difficult time trying to stay clean and healthy.

Portable battery-powered bucket shower

21

BUILDING YOUR OWN SOLAR POWER SUPPLY

MANY SMALLER ELECTRONIC DEVICES, INCLUDING YOUR cell phone, have their own internal rechargeable battery that can be recharged using a foldout solar module as described in chapter 5. However, for those devices having rechargeable batteries that cannot be removed, it may not be convenient to leave them sitting outside in the sun all afternoon.

There are also 12-volt DC devices we have discussed that do not have internal batteries, but could be powered from an external battery when the grid is down. These include medical equipment, flat-screen televisions, and DVD players. For these larger devices it is much simpler to provide backup power using a sealed AGM deep-cycle battery in a portable case and one or two separate solar modules in the 75- to 125-watt range.

During an extended power outage, you can temporarily mount the

Build your own portable solar power supply.

solar module on a homemade ground support using a few pressure-treated boards located in a sunny, south-facing area of your yard. Of course, you could prebuild this mounting rack and leave the module(s) locked up until disaster hits, if theft is a concern. A pair of wires could be unrolled and connected to the battery using a battery quick-disconnect.

Depending on the length of wire, the battery could be located either inside or under the solar array. At the end of the day, the modules could be moved back inside to prevent theft, and the battery could then be carried from room to room to power entertainment systems that do not have their own rechargeable batteries, or to recharge the smaller electronic devices that do.

Since most of the devices described in this text will have a charging adaptor to fit a 12-volt vehicle dashboard receptacle, any homemade 12-volt battery-backup system should include a cigarette lighter–type socket with in-line fuse. You can also attach a small 100- to 300-watt inverter to the battery box cover to power any small 120-volt AC devices that do not have 12-volt DC adaptors.

The following section will provide detailed directions for making your own portable backup battery system. Additionally, you could leave a small "trickle" charger connected to the battery and plugged into a wall outlet that would keep this battery fully charged until needed during a power outage. It's also possible to connect a higher amp capacity automotive battery charger if you need a quick charge using a generator during a power outage, which will minimize generator fuel usage.

Fuse and battery wiring to 12-volt socket

I suggest basing your design on a group 27- or 31-size deep-cycle 12-volt AGM battery which are reasonably priced but still able to handle repeated charging and discharging. These sealed batteries are not usually sold in automotive supply outlets, so try a boating supply store, as they are popular for powering electric trolling motors. An acid-proof and very rugged plastic battery case with a removable cover is also available and low cost. These are sold in multiple battery sizes, so pick the model designed to hold a group 27- or 31-size battery.

Next purchase a 12-volt DC dashboard utility outlet. These are sometimes sold with a cigarette lighter, which you can discard. Drill or cut a hole in the battery case cover to mount the 12-volt DC utility outlet. RV and boating stores sell a version of these outlets with a rugged flat mounting plate, which is much easier to install in the battery case cover and has a nice, professional look.

Using number 12 standard copper wire and an automotive-style in-line 15-amp DC-rated fuse, connect the "center" terminal of the utility outlet through the fuse to the positive terminal of the battery. Next connect the terminal on the outer shell of this utility outlet to the negative battery terminal. Use crimp-on ring terminals having a 3/8-inch hole, which should fit most battery posts with bolt type connection. If desired, you can install a small 100- to 300-watt DC-to-AC inverter on the top of the cover. These usually need free airflow around their cooling fin housings and may also include system status indicator lights, so the top mounting works best.

Completed system with optional 300-watt inverter

Many of the small DC-to-AC inverters include a set of interchangeable alligator clips and ring terminals. I recommend using the ring terminals, which should be bolted to the battery posts. Normally, any red cable is considered the positive terminal in automotive and boating wiring, and white or black is used to indicate the negative wire. Be aware that this is backwards from code-approved residential house wiring, which uses the black wire for "hot" and the white wire for "neutral."

You will need a solar charge controller rated for the amp output capacity of the solar module or modules. A 100-watt module will require a solar charger rated for at least 9 amps. Double this for 200 watts of solar array capacity. I prefer the solar charge controllers designed to install into a knock-out opening in the junction box on the back of the solar module.

However, if your solar module has a small, sealed junction box, there is room to mount this solar charge controller on either the inside of the battery box or on the cover next to the inverter. Do not attempt to electrically connect the solar module directly to the battery terminals without a solar charge controller, and do not buy a low-cost solar module. During an extended power outage you do not want to rely on a battery backup system consisting of inferior-quality components that could fail.

Module rear junction box with weatherproof solar charge controller

Chapter 5 provided more detail on the different types of batteries, solar modules, and solar charge controllers available, and the wiring diagram in the appendix shows how to wire all of the parts, so be sure to review this information before selecting the components you will need.

22

EMP PROTECTION OF BATTERY-POWERED DEVICES

WHILE NATURAL DISASTERS AND MAJOR STORMS can cause a week-long power outage in parts of the United States, a true grid-down event will most likely be caused by grid sabotage or an electromagnetic pulse (EMP). If you are living through a true grid-down period, the widespread power outage was probably caused by either a man-made EMP weapon or an extremely large solar flare. Regardless, most electronic devices having microprocessors and integrated circuit (IC) chips will be destroyed unless they were stored inside EMP shielding.

An electromagnetic pulse is not radioactive, and it does not harm people or animals. However, a nuclear detonation at least 250 miles above the central United States would blanket this country with a very damaging energy pulse that would cause high voltage and current to be induced into any type of electrical and electronic equipment wiring. Although similar to the effects of lightning on electrical systems, this induced high-voltage spike from an EMP is multiple times more powerful than a nearby lightning strike. Furthermore, the voltage induced by EMP in any electric wire ramps up to its peak level in a microfraction of a second, which is many thousands of times faster than a lighting strike.

Lightning surge suppressors used to protect computers cannot switch off fast enough to block this fast-moving high-voltage spike, so it will pass straight through as if there were no protection device.

While the EMP is not radioactive, if it was the result of a nuclear explosion, there is the potential for radioactive dust settling downwind, which can take a week or more to reach a safe level depending on the weather and how far away you live. A thorough discussion on radiation and its effects is beyond the scope of this book. However, there are several small, battery-powered radiation metering devices you may want to consider that are easy to operate and priced for the general public.

High-voltage transformers in substations can be destroyed by EMP.

Testing has shown that an EMP can destroy large distribution transformers. The power lines will serve as an antenna to collect and carry this energy back into electrical equipment not designed to handle these huge voltage spikes.

Since many of these large transformers are now manufactured in Europe and can take up to a year to build, ship, and install, it is understandable why this will be the most likely cause of a widespread grid-down event, and why it will take much longer to repair than downed power lines caused by a hurricane. While power lines, phone lines, and antennas can collect and transfer these damaging high-voltage effects down the line and into connected electrical devices, EMP does not rely solely on wires to enter this equipment. It travels through the atmosphere in every direction from the source, just like radio or television electromagnetic energy leaving a tall antenna.

EMP can easily pass through walls, roofs, windows, and any tiny opening or joints between metal-to-metal surfaces, although it cannot actually pass through metal.

Even when you are in a metal-enclosed elevator car in the center of a metal-framed building, your radio and cell phone will sometimes still receive very low-power cell phone and radio waves from distant radio towers, so you can imagine how this same type of energy magnified millions of times will have little problem reaching and destroying your unshielded electronic devices.

The induced currents an electromagnetic pulse generates in any electrical device not connected to power lines or antennas are typically only a few amps. This level of current will not damage the heavier wires in larger appliances, such as an electric heater or washing machine motor. However, these same induced currents at these same levels will easily fry the microscopic wiring inside any transistor, microchip, and integrated circuits typically used in the digital controls of the electric heater, washing machine, and every computer and electronic device found in today's homes and offices.

EMP testing has also shown similar damage to many of today's cars and trucks since they are full of electrical wires connecting multiple engine sensors. Many of today's vehicles contain up to a hundred microprocessor devices controlling everything. In 2004, Congress formed a commission to study the effects of EMP on the United States. This initial study was updated in April 2008 with an updated report titled *Report of the Commission to Assess the Threat to the United States for Electromagnetic-Pulse (EMP) Attack*. During this study, thirty-seven cars and eighteen trucks representing a cross-section of many different manufacturers and model years were separately placed in an EMP test chamber and tested both with engines running and engines off.

While there was little damage at lower levels of EMP radiation when the engines were off, at higher test levels 70 percent of these vehicles had minor repairable damage and 15 percent were never to run again. Most vehicles' engines stopped dead at higher levels of EMP, but many could be restarted. However, this would still be a disaster if you and everyone around you were traveling seventy miles per hour down the interstate and every car and truck stopped at the same time.

This study found it was not possible to predict which vehicles were more resistant to EMP damage, except for those models built before manufacturers switched over to electronic ignition systems in the early 1970s. Vehicle manufacturers learned during this early design transition that their first electronic ignition systems could be damaged by keying a nearby truck's CB radio transmitter, a lightning strike, or when driving past the radar dish at an airport.

Soon manufacturers started shielding the vehicles' wiring and improving the enclosures on all ignition and control modules to limit these interference problems. However, this minimum level of shielding cannot protect the sensitive electronics from the huge electromagnetic radiation produced by an EMP.

The military is finally waking up to the likelihood of losing most civilian communication systems from EMP damage, and has started to require manufacturers of military-grade electronics to provide better shielding. However, almost nothing has been done to provide better shielding of our national electric grid and the computer systems controlling everything from utility distribution to traffic control, subways, trains, airlines, radio stations, earth-orbiting satellites, television stations, and every electronic device you currently own. In other words, the only sure thing an individual can do to protect against EMP damage is to place these devices or their spares inside EMP shielding.

While there has been very little EMP testing of solar power systems, most of the battery wiring and batteries should be heavy enough to withstand the expected levels of current and voltage that are typically generated by an electromagnetic pulse. However, all solar charge controllers and inverters utilize microprocessors and integrated circuits in their construction, and these will most likely not survive an EMP event. This is especially true with the metal-framed solar modules on a roof serving as a giant antenna to collect the EMP energy and route it down through the wiring into these electronic devices. My advice is to have a spare charge controller and a backup inverter stored in a shielded enclosure.

Solar charge controller fried by lightning

An electromagnetic pulse should not create a high enough current and voltage to damage the electrical section of a typical generator, but almost all modern generators now include digital display panels and microprocessors to monitor the engine status and regulate generator output. Since these electronic components will not survive an EMP, the generator most likely will not run with this control system damage. If you have an expensive backup generator and a substantial supply of fuel, you may want to order a spare plug-in control board and store it in an EMP-proof container.

In July 1962, the United States government placed a 1.4 megaton nuclear bomb on a Thor missile, which was detonated 250 miles above Johnson Island in the Pacific Ocean. Although small compared to today's earthbound nuclear explosions, this "Starfish Prime" test is still the largest nuclear explosion ever detonated at high altitude. At this altitude the resulting electromagnetic pulse took out three hundred streetlights and set off numerous burglar alarms in Hawaii, almost one thousand miles away. In addition, the skies in the entire Southern Hemisphere were illuminated for many nights from the resulting aurora effect.

For comparison, the United States produced 940 MK-36 nuclear bombs before they were retired after 1962, and each had a yield over ten times the size of the Starfish Prime explosion. Just imagine what today's nuclear technology could do, especially since the state of electronics in the 1960s did not include any microchips or other integrated circuits easily destroyed by electromagnetic energy.

The majority of concern is an EMP generated by a nuclear bomb. However, not only does the sun also periodically emit EMP radiation, but under certain conditions it actually could do the same damage to all electrical systems in the United States as a nuclear detonation. The sun follows a fairly regular eleven-year sunspot cycle, and during most of the cycle, there are very few sunspots and limited magnetic radiation striking the earth. When the eleven-year cycle is at its peak, with lots of sunspots, gigantic "clouds" of charged particles are given off much like a dust cloud following a vehicle traveling down a dirt road.

Fortunately, most of the time when this occurs, the earth and the ejected particles do not cross paths, so there is little or no damage. When their paths do cross, these charged particles enter the upper atmosphere, producing damaging electromagnetic radiation, which spirals down to the earth's surface following the earth's magnetic lines. This high level of electromagnetic energy leaving the sun may affect only the half of the earth that was facing the sun when it happened, but that's still *half* of the earth.

In the 1800s, when there was very little electrical equipment in existence, the majority of wires between towns were telegraph lines. On

September 1, 1859, the earth was hit by a massive solar flare, which in turn induced high voltages in these cross-country wires. Once the voltage spike reached the end of the wires, several telegraph offices caught fire and many telegraph operators were shocked. Fortunately, the damage was limited since most of the electrical equipment and all of the modern electronic appliances were still years away from being invented.

This was called the "Carrington event" in recognition of Richard Carrington, an amateur astronomer in England who had been studying the sun's sunspot activity at the same time and tied the damage of the electrical equipment on earth to the massive solar flare he had just observed. Unfortunately, when this happens again—and it will—the damage to the electrical grid, computer networks, communication systems, and all earth-orbiting satellites could take years to fully recover. Don't wait to start your own emergency preparedness.

An old, discarded, and broken microwave oven will protect any electronic devices stored inside as long as the door seals are not damaged. New and never-used metal paint cans with tight fitting metal-to-metal lids also provide excellent EMP shielding for smaller electronic devices, as

5-gallon homemade EMP storage container

long as the stored devices are prevented from touching the metal interiors by plastic or cardboard. Metal ammo cans may appear to provide good EMP shielding, but they are actually a poor substitute since their lids have rubber gaskets. You must have a tight metal-to-metal fit between the lid and container of any enclosure you use to block EMP energy.

For under twenty-five dollars you can build your own EMP-proof container. I keep a battery-powered AM radio, shortwave radio, several FRS walkie-talkies, a small portable TV, a DVD player, and a GPS unit in a five-gallon plastic bucket, placed inside a six-gallon metal trash can with a tight-fitting metal lid. There is still room for extra rechargeable batteries and a fold-up solar battery charger. I filled the empty space between the walls of the inner bucket and the outer trash can with Crack insulation foam which is available in spray cans.

This homemade container keeps out rain, heat, cold, moisture, and chewing rodents, and as a bonus, it is EMP-proof. After the lid is closed tight, for added safety I wrap the joint with a layer of two-inch-wide metal-foil tape normally used to seal residential ductwork and duct insulation.

Adding metal-foil tape to seal lid joint

In closing this chapter, I would like to point out how all of the battery-powered appliances and electronic devices I have recommended to have during a grid-down event will easily fit in just two or three of these homemade five-gallon containers. This means all of your electronic devices would survive almost any EMP disaster, whether caused by a solar flare, an EMP weapon, or a high-altitude nuclear detonation. Since many of these battery-powered devices are not needed during normal, day-to-day living, storing them in these EMP- and moisture-proof metal containers is low-cost insurance to keep them safe until disaster strikes, and easy to grab and go if you have to bug out.

Larger equipment containing microelectronic circuits, including power inverters and audio/video equipment, can be protected by two separate layers of heavy-duty aluminum foil. This is usually easier to do if they stay in their original rectangular packaging and the foil is applied as if wrapping a birthday present. Again, metal-foil tape can be used to seal all joints and repair any tears.

In February 2016 North Korea test-fired a ballistic missile that passed directly over California on its path to earth orbit. Many in our government claim this test was a failure because the satellite did not appear to be working and may not have reached the intended orbit. However, this is little comfort since the payload sitting on top of their next rocket could be a nuclear EMP device and not a similar-sized communications satellite.

Keep in mind it does not require a ballistic missile having a nine-thousand-mile range following a trajectory directly over the United States to launch a satellite in an orbit just a few hundred miles above North Korea. Many feel this was actually a test of their long-range missile technology to learn how to place a nuclear EMP weapon 250 miles directly above the central United States. While we are being told the test was a failure, if North Korea were to place a "normal" satellite in orbit, this may be secondary to the country's real goal of testing how to use a long-range ballistic missile to launch a nuclear payload directly over the United States.

Other countries are not only developing EMP weapons, but their military systems are being "hardened" against EMP damage. At the same time, the United States has made little or no effort to harden our power grid and communication systems against natural or man-made EMP damage.

A congressional study titled *Securing the U.S. Electric Grid* was issued in July 2014 by the Center for the Study of the Presidency & Congress (CSPC), which was chaired by the Honorable Thomas F. McLarty III and the Honorable Thomas J. Ridge. The document's subtitle is *Understanding the Threats to the Most Critical of Critical Infrastructure, While Securing a Changing Grid*. While I think even I could have come up with a better title, this lengthy report does a thorough job of highlighting the many concerns I have also expressed in this text, and includes the testimony of many government and utility experts.

Although several years have passed since this report was issued, there have been no public statements by either the government or the major electric utilities that any of the recommendations made in this report have been started, let alone completed. While some secrecy is to be expected, you would assume any major renovations being made to thousands of power lines, substations, and generator plants spread across the entire country would be visible, or at least that a ton of new construction bids would be advertised. Sadly, I fear nothing of substance has been accomplished.

23

CLOSING COMMENTS

DO YOU SOMETIMES FEEL LIKE A ship sailing into the unknown? While nobody knows what the future holds for us, we can all be better prepared for the coming storm.

When you were a kid, I bet you had an uncle or aunt the rest of the family joked about as being odd or eccentric, always promoting ideas or a lifestyle everyone else thought was just plain loony. Well, sorry to be the one to break the news, but more than likely, if you are reading this book, you may have grown up to *be* that person in your extended family. In many cases, the hardest people to convince that it's time to prepare will be family members and friends.

For the first few years, the prepper community played the part of Paul Revere, only to be ridiculed for saying, "The sky is falling! The sky is falling," especially after Y2K didn't cause the end of the world. More and more preppers these days are realizing that a large part of our society will never believe this country is in for some real hard times and see no reason to prepare. I have overheard some say they will just take what they need if things get really bad.

I have also heard others say they will come to my home if things get bad, since they know we have prepared—but that's not going to happen! Many plan to just hunt when they need meat. Perhaps they do not realize that during the Great Depression there was not a deer, rabbit, or squirrel to be found. They didn't reappear until years later because everyone back then had the same idea. Many serious preppers today are keeping a much lower profile in their communities and have started to form loose-knit groups of like-minded individuals who can band together if things really turn ugly.

Most preppers will still go out of their way to help others who do not know how to begin. However, these days you may have to initiate the contact. Many preppers have found out the hard way how the nail sticking up gets hammered down. If you are just getting into prepping, I strongly recommend attending one of the many preparedness expos offered every few months in many parts of the country. You will be overwhelmed with the new products plus the free educational presentations.

It really makes no difference if the coming grid-down event will be caused by an EMP weapon, sunspots, aging infrastructure, natural disaster, sabotage, or incompetence. The result will be the same. Your state, and perhaps your entire section of the United States, could be without the electric grid for months, and long past your generator's fuel supply. Both Hurricane Katrina in New Orleans and Hurricane Sandy in Long Island proved government relief agencies are incapable of a rapid aid response and cannot provide enough emergency food, water, and shelter when millions of residents are affected.

There have already been multiple acts of sabotage to isolated utility substations, which are rarely reported by the national news media. These may be rehearsals for a future coordinated attack at multiple locations. With the rapidly changing world events, we can expect these acts to continue and most likely impact far larger sections of our utility infrastructure in the near future. Local electric utilities are constantly expanding the existing electric grid to add new homes and businesses.

However, a large number of the original substations and high-voltage transmission lines supplying these new loads from the distant generating plants were built in the 1960s and many are way past their design life. As mentioned in chapter 22, there have been numerous Senate hearings highlighting the vulnerability of the national electric grid to an EMP event, whether caused by terrorism or sunspots, yet little has been done to harden these systems against major destruction.

The steps outlined in this book will help you reduce your electric load when needed to stretch limited generator fuel supplies. I have also shown how you can utilize solar power to charge all types of battery-powered appliances and electronic devices after your generator runs out of fuel. When combined with other prepper books that address emergency food and water, home security, first aid, and weapons handling, this knowledge will help protect you and your family, regardless of what will cause the coming grid-down event or how long it will last.

Stay safe and good luck!

APPENDIX

USEFUL TABLES AND WIRING DIAGRAMS

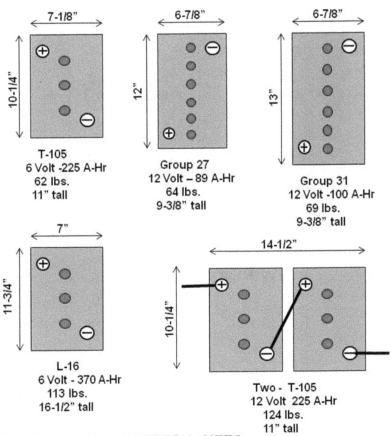

TABLE 1 - AGM BATTERY SIZES

APPENDIX

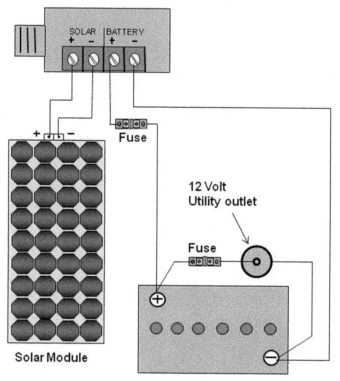

WIRING DIAGRAM

APPENDIX

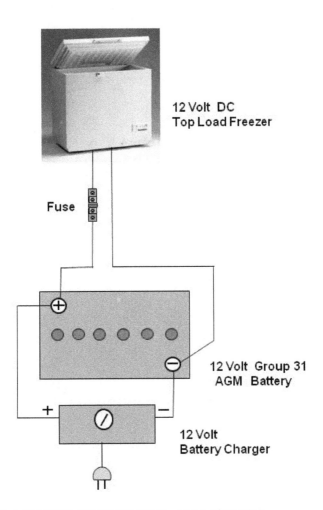

DC FREEZER WIRING DIAGRAM

TABLE 2 – 2% WIRE LOSS TABLE

	AMPS	#12	#10	#8	#6	#4	#2	1/0	2/0	4/0
12 VOLTS	5	14	22	36	57	91	144	230	290	461
	10		11	18	28	45	72	115	145	230
	15			12	19	30	48	76	96	153
	20				14	22	36	57	72	115
	40					11	18	28	36	57
	60						12	19	24	38
	100							11	14	23
24 VOLTS	5	28	45	72	114	182	289	460	580	923
	10	14	22	36	57	91	144	230	290	461
	15		15	24	38	60	96	153	193	307
	20		11	18	28	45	72	115	145	230
	40				14	22	36	57	72	115
	60					15	24	38	48	76
	100						14	23	29	46
	175							13	16	26
48 VOLTS	5	56	90	144	229	364	579	921	1161	
	10	28	45	72	114	182	289	460	580	923
	15	19	30	48	76	121	193	307	387	615
	20		22	36	57	91	144	230	290	461
	40		18	18	28	45	72	115	145	230
	60		15	12	19	30	48	76	96	153
	100					18	28	46	58	92
	175							26	33	52

TABLE 2 lists the maximum distance between two wiring components for various combinations of wire size, current, and voltage, to maintain a 2 percent voltage drop. You should only count this one-way distance since the resistance for the total length of wire out and back in the circuit is accounted for in these table values.

TABLE 3 – MAXIMUM WIRE AMPERAGE RECOMMENDATIONS

WIRE GAUGE	BATTERY & INVERTER WIRING 30°C (86°F) AMBIENT NOT IN CONDUIT THHW	BATTERY & INVERTER WIRING 30°C (86°F) AMBIENT IN CONDUIT THHW	UNDERGROUND DIRECT BURIAL 30°C (86°F) AMBIENT UF	EXTERIOR SOLAR 80°C (176°F) AMBIENT IN CONDUIT THWN-2, USE -2
# 12 GAUGE	20	20	20	8
# 10 GAUGE	30	30	30	12
# 8 GAUGE	70	50	40	22
# 6 GAUGE	95	65	55	30
# 4 GAUGE	125	85	70	38
# 2 GAUGE	170	115	95	53
# 1 GAUGE	195	130	110	61
# 1/0	230	150	125	69
# 2/0	265	175	145	79
# 3/0	310	200	165	92
# 4/0	360	230	195	106

UNDERSTANDING WATTS, AMPS, AND VOLTS

It's easier to understand an electrical circuit if you think of a plumbing system. Think of the storage battery as a water tank full of water, and the higher the water level (higher voltage) in the tank, the higher the pressure. The electricity flowing through the wires is represented by the water flowing through the pipes.

To see how much pressure is pushing the water through the pipes, we use a "pressure" gauge. In an electric circuit, we measure the pressure pushing the electricity through the wires as "voltage" using a voltmeter. The battery voltage is an indication of how much pressure it has, which tells us the charge level. Table 4 at the end of this appendix can be used to convert battery voltage into a percent of charge level for a standard lead-acid deep-cycle battery.

To measure how much water is flowing in the plumbing pipes at any point in time, we use a flow meter to measure the gallons or rate of flow per second, minute, or hour. In an electric circuit we measure the rate of electrical flow through the wires in "amps" using an amp meter.

Energy is a measurement of work defined as a force over a period of time. For example, we say a car traveled sixty miles. However, if the car traveled sixty miles in one hour, we say its rate of speed was sixty miles per hour. At a rate of sixty miles per hour, the car will not actually travel sixty miles unless it stays at this speed for the full hour.

Electrical power is measured in watts, with 1000 watts called a kilowatt (kW). However, this is the rate power is flowing through an electrical device at any instant of time. To know how much "energy" is being consumed or produced, we need to include time to convert this rate measurement into energy, or kilowatt-hours (kWh). Adding the time factor "hour" converts the rate at which this power is being used or produced into power. For example, if a toaster oven has a nameplate that indicates it draws 1000 watts, or 1 kW, this is its "rate" of electricity usage. We need to add time to measure the actual energy that was consumed, billed, and generated. This means our toaster oven consumed 1000 watt-hours of power, or 1 kWh if it operated one full hour. If we let it run for two hours, it consumed 2 kWh, not 2 kW. Your house may be drawing 8 kW of power from the grid right now, but your electric meter has to record the amount of time it was using this level of energy plus add all the other levels over an entire month to determine how many kilowatt-hours (kWh) to bill.

In our plumbing system, pressure and flow are interrelated, since the more pressure you have, the more flow you get. Also related to plumbing, a higher pressure can force more flow through a smaller pipe, but a larger pipe can pass more flow without increasing the pressure. In an electric circuit, a larger wire size can carry more amps with less voltage loss.

The great thing about calculating the amps, volts, or watts of energy flowing through any DC electrical circuit is if you know two of these values, it's easy to find the third term, which you can almost calculate in your head.

$W = A \times V$ where "W" is watts, "A" is amps, and "V" is volts

Of course you can switch the terms around and get:

$A = W / V$ and $V = W / A$

EXAMPLE 1: You are planning to power a 12-volt DC deep-well pump rated at 300 watts with a 12-volt RV battery, and want to know the amp flow to size the wire required to make the connection. The math would be:

$A = W / V = 300$ watts $/ 12$ volts $= 25$ amps

Most calculated "amp" values in an electrical circuit are usually required by code to have a 25 percent added safety factor or a 1.25 percent multiplier, which is $25 \times 1.25 = 31.2$ amps, so according to table 3, a no. 8 wire rated for 40 amps in a direct-bury application will work assuming the wire to the well is buried. However, we also need to check table 2 to make sure the length of this wire will not have a high voltage drop. In this example, Table 2 indicates a no. 8 wire supplying a 12-volt load is not recommended, so you will need a much larger wire to limit the voltage drop, or the pump may not have enough voltage to run. This is why most battery-based systems supplying larger pumps, motors, and inverters use a higher, 24- or 48-volt battery bank.

Now let's see how long the battery can power this pump load. You purchased a 12-volt battery with a 96 amp-hour capacity, but we always want to avoid discharging any battery below 50 percent since repeated deep-discharging will significantly shorten battery life. This means we have 48 amp-hours (96×0.50) of battery capacity to work with, so 48 amp-hours useful storage capacity means we can power this pump slightly less than two hours. Obviously we will need a larger battery!

48 amp-hours $/ 25$ amps $= 1.9$ hours

Remember: calculating power for any load always includes a time factor.

SIZING SOLAR ARRAY WIRE

Sizing any wire connecting a solar module is slightly different from all other wire calculations. This includes the wire between the solar module(s), the wire from the solar array to any fuse or combiner box, and the wire to the solar charge controller. Normally a safety factor of 25 percent is added to the "short circuit" amp rating when calculating wire sizes.

However, unlike a lightbulb or electrical appliance, both of which have their maximum wattage and amp ratings clearly listed on the nameplate, under extremely low temperature conditions or increased solar intensity due to reflected snow, the output of a solar module can briefly exceed its published nameplate ratings. For this reason, testing agencies require adding an additional 25 percent factor to any solar array wiring.

EXAMPLE 2: What wire size is required to connect four (4) solar modules wired in series to a charge controller forty feet away?

The nameplate indicates each solar module is rated at 5.6 amps at 12-volt load, and has a 6.3-amp short circuit rating.

(6.3) (1.56) = 9.83 amps at 48 volts.

Always use the "short circuit" amp rating of any solar module to calculate wire size for array wiring back to a combiner box or charge controller and add the second 25 percent multiplier (1.25 × 1.25 = 1.56 percent).

From table 2, a no. 10 wire will carry 10 amps up to 45 feet at 48 volts.

TABLE 3 indicates a no. 10 type "USE-2" wire is approved for up to 12 amps at 176 degrees Fahrenheit, which is the answer for this application. The wire selection for any solar module wiring needs to be rated for a higher temperature since a solar module can easily reach some very high temperatures lying in the sun all day. Solar wiring should also have an insulation that is both sunlight and water resistant, which is why most solar installers choose the "USE" wire, which has these characteristics. The "-2" added to any wire designation indicates a wire that has a very high temperature rating.

INVERTER WIRE SIZING

Several places in this text describe the amps and volts going into and out of an inverter and provide values, but do not show how these values were calculated.

In simple terms, the energy going into any inverter must equal the energy coming out of the inverter. Of course, we are simplifying again by not including the efficiency losses inside the inverter which become heat energy. This efficiency loss will actually be in the lower 4 to 6 percent range for a high-quality inverter operating near its design load, and 15 to 20 percent for lower-quality inverters. Not counting this efficiency loss, to keep it simple, we get:

ENERGY IN = ENERGY OUT

From our previous example #1, we learned that energy, or watts = amps × volts, and amps = watts / volts.

EXAMPLE 3 - For an inverter connected to a 12-volt battery, what is the amp load on the battery and battery wiring when the inverter is powering a 2400-watt load at 120 volts AC?

Amps = watts / volts, so: 2400-watts / 120 volts = 20-amp flow *out* of the inverter to supply a 2400-watt load.

However, if we get 2400 watts out of an inverter, we need to put at least 2400 watts of energy into the inverter, plus enough extra to cover the additional efficiency losses.

Amps = watts / volts, so 2400 watts / 12 volts = 200-amp flow *into* the inverter from the battery!

Keep in mind this value does not include the inverter's efficiency losses, which will push the amp flow even higher than the 200-amp required flow in the battery cables from a 12-volt battery to satisfy the output load.

If this were a 24-volt DC inverter and battery, the amps would be cut in half, or 100 amps for the same 2400-watt load:

2400 / 24 = 100 amps

For a 48-volt inverter, the 48-volt battery amp load would be 50 amps.

2400 / 48 = 50 amps

This is why most portable and vehicle inverters that must use a 12-volt vehicle battery rarely exceed 1200 watts capacity, which would be a 100-amp load on the batteries and battery cables.

1200 watts / 12 volts = 100 amps

Using table 3 for a 100-amp load at 12 volts, you would need a no. 4 wire to connect the inverter to the car battery, which is larger than most car jumper cables! In addition, most light-duty car batteries could not take a constant 100-amp load, and most car alternators could not offset this much battery load.

TABLE 4 – BATTERY STATE OF CHARGE (VOLTS)

% FULL CHARGE	HYDROMETER READING @ 77° F	12-VOLT BANK	24-VOLT BANK	48-VOLT BANK
100%	1.266	12.64	25.27	50.54
95%	1.259	12.59	25.18	50.36
90%	1.251	12.55	25.09	50.18
85%	1.244	12.50	25.00	50.00
80%	1.236	12.46	24.91	49.82
75%	1.229	12.41	24.82	49.64
70%	1.221	12.37	24.73	49.46
65%	1.214	12.32	24.64	49.28
60%	1.206	12.28	24.55	49.10
MAXIMUM DAILY DEPTH OF DISCHARGE				
55%	1.199	12.23	24.46	48.92
50%	1.191	12.19	24.37	48.74
45%	1.184	12.14	24.28	48.56
40%	1.176	12.10	24.19	48.38
NEVER EXCEED DEPTH OF DISCHARGE				
35%	1.169	12.05	24.10	48.20
30%	1.161	12.01	24.01	48.02
25%	1.154	11.96	23.92	47.83

The above table is for battery cells at 77° F (27° C) and measured when not under load or charging.

The following temperature corrections should be applied to Table 4 voltage readings when ambient temperature is above or below 77° F.

AMBIENT TEMP.	12 VOLT	24 VOLT	48 VOLT
95° F / 32° C	- 0.3 V	- 0.6 V	- 1.2 V
85° F / 29° C	- 0.1 V	- 0.3 V	- 0.5 V
65° F / 18° C	+ 0.2 V	+ 0.4 V	+ 0.8 V
55° F / 10° C	+ 0.4 V	+ 0.7 V	+ 1.5 V
45° F / 7° C	+ 0.5 V	+ 1.1 V	+ 2.1 V
35° F / 2° C	+ 0.7 V	+ 1.4 V	+ 2.8 V

Repeated discharge below maximum levels indicated will significantly shorten battery life. In high ambient temperature locations, a more diluted electrolyte may be required and different hydrometer readings will result.

REFERENCES

Out of an endless sea of books on all things off-grid and solar, I think you will find my personal picks to be excellent sources for more detailed information on these topics and great additions to your reference library.

BOOKS

Lester Brown, *Plan B 3.0: Mobilizing to Save Civilization*, rev. and exp. (New York: W. W. Norton, 2008).

Center for the Study of the Presidency & Congress (CSPC), *Securing the U.S. Electrical Grid: Understanding the Threats to the Most Critical of Critical Infrastructure, While Securing a Changing Grid* (Washington, D.C.: CSPC, 2014), https://www.thepresidency.org/sites/default/files/Grid%20Report%20July%2015%20First%20Edition.pdf.

Joel Davidson and Fran Orner, *The New Solar Electric Home: The Complete Guide to Photovoltaics for Your Home* 3rd ed. (n.p.: Aatec, 2008).

REFERENCES

EMP Commission, *Report of the Commission to Assess the Threat to the United States from Electromagnetic Pulse (EMP) Attack* (2008), http://www.empcommission.org/docs/A2473-EMP_Commission-7MB.pdf.

Ted Koppel, *Lights Out: A Cyberattack, A Nation Unprepared, Surviving the Aftermath* (New York: Crown, 2015).

MAGAZINES

Backwoods Home Magazine (www.backwoodshome.com)

Self-Reliance magazine (http://www.self-reliance.com/)

Home Power magazine (http://www.homepower.com)

MORE INFORMATION ON CHARLES KETTERING

Thomas Alvin Boyd, *Charles F. Kettering: A Biography* (n.p.: Beard Books, 2002).

"The American Wind Charger Industry: 1916 to 1960," http://www.windcharger.org/Wind_Charger/Welcome.html.

"Delco-Light Farm Electric Plant," http://www.doctordelco.com/Dr._Delco/Charles_Kettering.html.

RESOURCES

LIGHTING
LED Flashlights
    Fenix RC11 — www.fenixlighting.com
    Dorcy Metal Gear 200 — www.opticsplanet.com
    Pelican ProGear 3310PL — www.brightguy.com
LED Lanterns
    AYL Starlight — www.ledlantern.org
    Etekcity 1 Pack — www.ledlantern.org
    AGPtek 5 Mode — www.superbrightleds.com
    Goal Zero Lighthouse — www.thereadystore.com
    Mpowered Luci — www.mpowerd.com/
    Black Diamond Voyager — www.rei.com
LED Headlamps
    Black Diamond Revolt — www.blackdiamondequipment.com
    Black Diamond Spot — www.blackdiamondequipment.com
    Tec APEX — www.brightguy.com
    Coast HL7 — www.coastportland.com
LED Cabin Lighting
    M4 Products — www.m4products.com
    Brilliant Light — www.makariosrv.com
    Progressive Dynamics — www.rvupgradestore.com
LED Solar Walk Lights
    Moonrays Richmond — www.moonrays.com
    1x Large Super Bright — www.landscapeandlighting.net

RESOURCES

LED Solar Floodlights
    Sunforce 150 — www.ledsolarfloodlights.com
    Maxsa Solar 150 — www.allmodern.com
    Commercial Grade — www.outdoorsolarstore.com

RADIOS
All Band and AM/FM
    C Crane Skywave — www.ccrane.com
    Sony AMZICFM260AA — www.amazon.com
    Sangean DT-210 pocket — www.amazon.com
FRS/GMRS
    Motorola MT350R FRS — www.factoryoutletstore.com
    Midland LXT600VP3 — www.amazon.com
    Uniden GMR5089-2CKHS — www.buytwowayradios.com
2-Meter Ham
    BaoFeng BF-F8HP — www.amazon.com
    Yaesu FT-2900R 75-Watt — www.universal-radio.com
    Yaesu FT-270R 5-Watt — www.amazon.com
    YAESU FT-60R — www.hamradio.com
CB Radio
    Midland 75-785 40-Channel — www.amazon.com
    Cobra HH 38 WX ST 4-Watt — www.walcottradio.com
WEATHER
    Midland HH54VP Portable — www.amazon.com
    Oregon Scientific WR602 — www.smithgear.com

REFRIGERATORS / FREEZERS
Top Load
    SunDanzer 50 — www.sundanzer.com
    Unique 80 — www.uniqueoffgrid.com
Upright
    Unique 290 — www.uniqueoffgrid.com
    SunFrost RF12 — www.sunfrost.com
    Domestic CR-180 — www.altern-energy.com

WATER / WELL PUMPS
Deep Well
    Sun Pumps SDS-D-228 — www.sunpumps.com
    SHURflo 9300 — www.thesolarstore.com
    Aquatec SWP-4000 — www.thesolarstore.com
In-Line Pressure
    Flowlight Booster — www.dankoffsolarpumps.com
    Uquatec 5800 Series — www.aquatec.com
Hand-Operated
    Bison DW Standard — www.bisonpumps.com
    Hand Well Pump — www.preparewise.com
    Stainless Steel Pump — www.sunshineworks.com
Portable Shower
    Pure-Clean #PCSHPT12 — www.walmart.com
    Human Shower Ease — www.human-creations.com
    Caravan 12-Volt Washer — www.amazon.com

RESOURCES

BATTERY CHARGERS
Multiflashlight
    T9688 Super Universal   www.megabatteries.com
    AccuManager 16   www.amazon.com
    Knox 16-Bay Rapid   www.amazon.com
    Ansmann ENERGY 16   www.amazon.com
Battery Tools
    Dewalt 20-Volt #DCB119   www.lowes.com
    Dewalt 18-Volt #DC9319   www.lowes.com
    Ryobi #P131 In-Vehicle   www.homedepot.com
    Makita # DC18SE   www.homedepot.com
Solar for cell phone
    ZZero 8000   www.solarpowerbackpacks.com
    Brolar 5000   www.solarproductswarehouse.com
    Poweradd Apollo Pro   www.solarchargeroutlet.com
Solar for laptop
    Goal Zero Nomad 7 Plus   www.rei.com
    XDPOWERS 16W Folding   www.solarpowerbackpacks.com
    Aukey 21W Dual   www.rvsuppliesoutlet.com
Solar for backpacks
    Voltaic Systems Fuse 10   www.voltaicsystems.com
    BirkSun Elevate   www.solarpowerbackpacks.com

RECHARGEABLE BATTERIES
"AA" size
    Eneloop Pro   www.amazon.com
    Tenergy 2600   www.amazon.com
Deep Cycle – Open
    Trojan Golf Cart   www.trojanbattery.com
    Daka Golf Cart   www.eastpennmanufacturing.com
    Exide Golf Cart   www.exide.com
Deep Cycle – AGM/GEL
    Deka 8A31DT   www.thesolarstore.com
    Trojan 31 AGM   www.altestore.com

SOLAR SYSTEM HARDWARE
Charge Controllers – small
    Morningstar SunSaver10   www.wholesalesolar.com
    Esky 30A LCD   www.amazon.com
    Sunforce 60032 30-Amp   www.amazon.com
Charge Controllers – MPPT
    Outback Flexmax 60   www.invertersupply.com
    Morningstar Tristar 60   www.thesolarstore.com
    MidNite Solar Classic 150   www.wholesalesolar.com
Solar Modules – fold-up
    Nomad 20   www.goalzero.com
    SunJack 20   www.amazon.com
    Eco-Worthy 100W 12V   www.eco-worthy.com
Solar Modules – roof/ground

RESOURCES

125-Watt Starter Kit	www.mrsolar.com
Renogy 100-Watt Kit	www.amazon.com
Renogy Eclipse – 100-Watt	www.renogy.com
Inverters – portable	
400 W 12-Volt Inverter	www.amazon.com
Aims 300 W sinewave	www.theinverterstore.com
Inverter/charger – Off-Grid	
Aims 3000 W	www.theinverterstore.com
Outback 3648	www.outbackpower.com
Xantrex Freedom SW	www.xantrex.com
DC fuses and Safety	
Solar fuses and bkrs.	www.solar-electric.com
Solar fuses and bkrs.	www.backwoodssolar.com

SECURITY
Motion Sensors
    1byone Wireless Alert — www.amazon.com
    Doberman hanging alarm — www.amazon.com
Hidden Cameras
    Bushnell HD Essential E2 — www.basspro.com
    Cuddeback E2 Long Range — www.basspro.com

ENTERTAINMENT
Musical Instruments
    Yamaha YPG-235 76-key — www.sweetwater.com
    Casio CTK-2400 Portable — www.zzounds.com
Portable Amplifiers
    Roland KC – 110-Amp — www.guitarcenter.com
    Denon ENVOI 360w PA — www.agiprodj.com
    Roland CUBE Guitar Amp — www.musiciansfriend.com
    Roland BA330 Portable PA — www.musiciansbuy.com

NONPOWERED OFF-GRID DEVICES
COOKSTOVES – wood/charcoal
Portable
    Volcano 3-Fuel Portable — www. volcanogrills.com
    Esbit Folding Charcoal Grill — www.amazon.com
Kitchen
    Wood Stove Forum — www.cookstoves.net
    Obadiah's Wood Stoves — www.woodstoves.net
    Pioneer Princess — www.antiquestoves.com

WATER FILTERS
Individual
    Alexapure personal filter — www.mypatriotsupply.com
    LifeStraw 2-Stage — www.lifestraw.com
Tabletop
    Big Berkey — www.berkeyfilters.com

CLOTHES WASHING
Hand-Powered

RESOURCES

    Hand-Operated Agitator   www.beprepared.com
    Lehman's Hand Washer   www.lehmans.com
Wringers
    Calliger Crank Wringer   www.amazon.com
    Best Hand Clothes Wringer   www.amazon.com

WATERLESS TOILETS
Compost
    Excel-NE   www.sun-mar.com
    Envirolet DC12   www.envirolet.com
    Nature's Head   www.natureshead.net

YARD MAINTENANCE
Mowers
    Worx 56 V 19"mower   www.worx.com
    Neuton CE6 Cordless   www.neutonpower.com
    Stihl RMA 410 C   www.stihlusa.com
Trimmers
    Worx GT 2.0   www.worx.com
    Oregon ST275-E6   www.stringtrimmersdirect.com
Chain Saws
    Stihl MSA 160 C-BQ   www.stihlusa.com
    Husqvarna T536Li   www.husqvarna.com
    Makita XCU02Z 18V X2   www.amazon.com
    OREGON MAX CS300-A6   www.chainsawsdirect.com

EMP PROTECTION
Radiation Detectors
    NukAlert Keychain   www.nukalert.com
    Mini Rad-DX   www.laurussystems.com
EMP Shielding
    Shielding Bags   www.techprotectbag.com
    EMP Shielding Materials   www.sunshineworks.com

PREPAREDNESS SUPPLIERS
    Backwoods Solar Inc.   www.backwoodssolar.com
    Lehmans Supply   www.lehmans.com
    Real Goods Outlet   www.realgoods.com
    My Patriot's Supply   www.mypatriotsupply.com
    Emergency Essentials   www.beprepared.com
    Prepare Wise   www.preparewise.com
    American Survival Wholesale   www.americansurvivalwholesale.com

PREPAREDNESS EXPO SCHEDULES
    Sustainable Preparedness EXPO   www.susprepexpo.com
    Prepper Shows   www.preppershowsusa.com

PREPAREDNESS GROUPS & PODCASTS
    American Prepper Network   www.americanpreppersnetwork.com
    The Survival Podcast   www.thesurvivalpodcast.com
    Off the Grid News   www.offthegridnews.com

ABOUT THE AUTHOR

Jeff Yago was raised in Texas and experienced firsthand the destruction of major hurricanes hitting his town, which required evacuation to emergency shelters for days at a time. After graduating with degrees in mechanical engineering and industrial management, he began as a design engineer for a manufacturer of battery-powered industrial equipment.

He later served twelve years as energy director with several professional engineering firms before starting his own firm in 1990. He has completed solar and energy reduction projects for schools, hospitals, and military facilities throughout the United States and Europe. In addition to being a licensed professional engineer and certified energy manager, he is a NABCEP-certified solar professional, a licensed electrician, and a ham radio operator. His first book, *Achieving Energy Independence—One Step at a Time*, was published in 1998 and has since seen three reprints. His numerous how-to articles on solar- and battery-power topics have appeared in *Home Power* magazine, *Backwoods Home Magazine*,

ABOUT THE AUTHOR

Self-Reliance magazine, *Mother Earth News, Energy Engineering, Energy Users News, Countryside & Small Stock Journal Magazine,* and many other publications.

During his career Jeff Yago has been awarded five United States patents for his design work, was inducted into the Order of the Engineer for lifelong commitment to the engineering profession, and holds a private and commercial helicopter pilot's license. Finally, he is an accomplished keyboardist, playing with several local bands, and serves as substitute organist for his church.

Jeff and his wife, Sharon, live in Gum Spring, Virginia, in the solar home they built together in 1994, where they enjoy visits from three adult children and six grandchildren. When not writing, Jeff Yago teaches monthly solar and preparedness classes, and is an occasional guest lecturer at several area universities.

Jeff's website: www.offgridprepper.com
Follow Jeff on Facebook at Facebook.com/JeffreyRYago

NOTES

1. Center for the Study of the Presidency & Congress (CSPC), *Securing the U.S. Electrical Grid: Understanding the Threats to the Most Critical of Critical Infrastructure, While Securing a Changing Grid* (Washington, D.C.: CSPC, 2014), https://www.thepresidency.org/sites/default/files/Grid%20Report%20July%2015%20First%20Edition.pdf.
2. CSPC, *Securing the U.S. Electrical Grid*, 27–28.
3. Ibid., 29.
4. Ibid., 30–31.
5. Dick Brass, "Lightning strikes cause blackout in 1977," *New York Daily News*, July 14, 1977, http://www.nydailynews.com/new-york/lightning-strikes-blackout-1977-article-1.2284170.
6. Ohio History Central "2003 Blackout," http://ohiohistorycentral.org/w/2003_Blackout?rec=1653.
7. CSPC, *Securing the U.S. Electrical Grid*, 25–27.
8. Eric Roston and Blacki Migliozzi, "Obama's EPA Rule Is Redrawing the U.S. Coal Map," *Bloomberg*, April 13, 2015, http://www.bloomberg.com/graphics/2015-coal-plants/.

INDEX

A

AccuManager 20 charger (by AccuPower), 47
AC/DC power explained, 38–40
adaptors, 48, 50, 90, 145, 147, 149–52, 155, 156, 169
AGM battery sizes (table 1), 185
American Radio Emergency Service (ARES), 85, 153, 154
AM/FM radios, 78–79
alarms (security), 124–26
amperage recommendations, maximum wire (table 3), 189
amps, understanding watts, volts, and, 190–92
Ansmann Deluxe Energy 8 charger, 47–48
anthologies, storing on CD, 92, 93
attacks, utility grid attacks, 13–17

B

Backwoods Home Magazine, xi, xii, 200, 206
bathing, 166
batteries
    common household battery sizes, 44 (chart)
    main BCI groups and their amperage per hour rating, 63 (chart)
    most popular rechargeable, 44
    steps to making the change to rechargeable, 42–43
Battery Council International (BCI), 62
battery power
    bugging out with, 143–46
    clean water with, 110–17
    computers with, 89–94
    connecting devices to, 147–56
    emergency communication with, 77–88
    entertainment with, 95–99
    introduction to, 38–51
    lighting with, 71–76
    medical equipment with, 100–4
    portable tools with, 105–9
    refrigeration with, 118–22
    RV camping with, 138–42
    security with, 123–33
    using vehicles for, 134–37
battery-powered clocks, 132
battery-powered tools, first two to collect, 106
battery sizes (table 1), 185
Berkey water filter, 115
BiPAP unit, 100–2
BlackEnergy (malware virus), 13
books (recommended), 199–200
bridges, 15, 94
bugging out with battery power, 143–46
bugout bag, building a, 144–46
bugout vehicle, best, 139

C

cadmium, 42
cameras (security), 127–29
Canadian Standards Association (CSA), 57
Carrington, Richard, 178
Carrington event, 177–78
CB radios, 83–85
CDs, collecting reference books and other pertinent information on, 92, 93
cell phone charger, xiv, 43, 145, 150
cell phones, 80–81

INDEX

chargers, recommended, 47–48
chemistry of the various rechargeable battery types (chart), 45
citizen band radios. *See* CB radios
clean water with battery power, 110–17
clocks, battery-powered, 132
coal-fired power plants, percentage of power generated in the United States from, 24
clothing care. *See* washing without grid power
computer hacking, xi, 4, 9, 11, 12, 13–14, 17
computers with battery power, 89–94
    easiest way to keep a laptop or tablet charged without a generator, 90
computer virus, 12, 13–14
communication with battery power (emergency), 77–88
connecting devices to battery power, 147–56
cooking without grid power, 157–60
CPAP machine, 100, 102–3
cyberattack on Ukraine's power grid, 13

D

Dayton Engineering Laboratories, 26. *See* Kettering, Charles F.
DC (direct current). *See* AC/DC power explained
DC freezer wiring diagram, 187
Delco-Light plant(s), 25–31
devices
    connecting to battery power, 147–56
    EMP protection of battery-powered, 172–81
diagram
    DC freezer wiring, 187
    wiring, 186
Domestic Engineering Company, 26. *See* Kettering, Charles F.
dry cell batteries, estimated number discarded yearly in the United States, 42
DuPont Chemical Company, 29
DVD players and TV, 95–98

E

electrical utility infrastructure, highest risk threatening today's, 9
electromagnetic pulse. *See* EMP
emergency communication with battery power, 77–88
EMP, xi, 4, 7, 17–18, 36, 48, 81, 84, 156, 158, 172–81, 183, 184
    protection of battery-powered devices, 172–81
    resources on EMP protection, 205
    shielding, 84, 172, 175–76, 178–79
Energy Star efficiency rating, 49
entertainment with battery power, 95–99

F

Family Radio Service (FRS) walkie-talkies, 82
flashlights, 74–75
Freon, 29
Frigidaire (history of the), 29–30
fuses, wire and, 67–70

G

General Motors (GM), 27
generator
    easiest way to keep a laptop or tablet charged without a, 90
    life after, 32–37
grid-down event, 5, 7, 8–9, 20, 22–24, 36, 41, 42, 48, 49, 52, 57, 72, 76, 78, 81, 83–84, 85, 90, 93, 95, 103, 105, 107, 115, 118–22, 124–26, 129, 131, 132–33, 143–44, 146, 149, 155, 158, 159, 164, 165, 172, 180, 183, 184
    defined, 8–9
grid power
    cooking without, 157–60
    washing without, 161–66

H

hacker(s), 4, 11–14
hacking, xi, 4, 9, 11, 12, 13–14, 17
ham radio, 85–88, 206
hand pumps, 114
Hurricane Katrina, Hurricane Sandy, 9, 183

I

Internet, xiii, 2, 6, 11, 14, 77, 79, 81, 89, 90, 93–94, 95, 98, 121, 123, 132, 139, 162, 163
Internet routers, 2, 17, 22, 77, 90, 92, 93
inverter
    selecting an, 64–66
    wire sizing, 195–96
Israel, computer hacking in, 13–14

J

Jacobs wind generators, 30–31

INDEX

K
Kettering, Charles F., 25–31, 200

L
laptops. *See* computers with battery power
laundry. *See* washing without grid power
LED lanterns, 75–76
life in the early 1900s, 18–19, 25
light fixtures, 72–73
lighting with battery power, 71–76
lightning, 20, 22, 172–73, 175, 176
looting, 21

M
magazines (recommended), 200
Maha Ultimate Professional charger, 48
malware. *See* virus (computer)
maximum-power-point tracking (MPPT) charge controller, 59–60
McLarty, III, Thomas F., 181
medical equipment with battery power, 100–4
Memorial Sloan Kettering Cancer Center, 31
mercury, 41–42, 115
Metcalf sniper attack, 15
Midland HH54 portable weather radio, 131
MPPT charge controller, 59–60
musical instruments, 98–99

N
National Electric Code, xvi, 67, 70
National Emergency Broadcast System, 129
National Weather Service (NWS) radio(s), 129–31
New York City blackout of 1977, 20–21
North Korea ballistic missile test (2016), 180
NWS radios, 129–31

O
Oregon Scientific WR602 portable weather radio, 131

P–Q
Panasonic rechargeable batteries, 47
portable shower, 166
portable tools with battery power, 105–9
power, percentage generated in the United States from coal-fired power plants, 24
Powerex MH-C800S charger, 48

power lines, xiii, xiv, 10, 12, 13, 16, 174, 181
power tools. *See* portable tools with battery power
preppers, xii, 48, 86, 87, 137, 182, 183
pump controllers, 113

R
Radio Amateur Civil Emergency Services (RACES), 85, 153, 154
radio licenses, 86–87
radios. *See* AM/FM radios; CB radios; NWS radios; shortwave; two-meter band radios; weather radios
rechargeable batteries
  advantages and disadvantages of the various types of, 45 (chart)
  highest-rated, 47
  most popular, 44
  steps to making the change to, 42–43
reference books, storing on CDs, 92, 93
refrigeration with battery power, 118–22
refrigerators and freezers, good brands of battery-operated, 121
Report of the Commission to Assess the Threat to the United States for Electromagnetic-Pulse (EMP) Attack, 175, 200
Ridge, Thomas J., 181
Rural Electrification Act, 24, 26, 31
RV camping with battery power, 138–42

S
Sanyo "Eneloop" rechargeable batteries, 47
SCADA (supervisory control and data acquisition), 10–12
Securing the U.S. Electric Grid (congressional study) 14, 181
security with battery power, 123–33
  alarms, 124–26
  security cameras, 127–29
  security lighting, 126–27
self-help articles, downloading to CD, 93
shortwave
  radio bands, popular, 80 (chart)
  radios, 79–80
shower, portable, 166
sizing
  inverter wire, 195–98
  solar array wire, 193–94
Sloan, Alfred P., 31

INDEX

smoke detectors, 46, 131–32, 133
solar array wire, sizing, 193–94
solar battery, selecting a, 60–63
solar charge controller, selecting a, 58–60
solar module
    label explanations, 56 (chart)
    selecting a, 53–56
        three major types available to the general public, 55
solar pathway lights, 73–74
solar power supply, building your own, 167–71
solar system components, understanding, 52–70
Starfish Prime, 177
substations, 10, 11, 12, 13, 14–16, 39, 17, 181, 183–84
supervisory control and data acquisition. *See* SCADA
Survival Podcast, The, xiii–xiv, 205

T

tables
    AGM Battery Sizes (table 1), 185
    Battery State of Charge (Volts) (table 4), 197
    Maximum Wire Amperage Recommendations (table 3), 189
    2 Percent Wire Loss Table (table 2), 188
temperature corrections (battery voltage), 197–98
terminals, preferred method to connect larger-size battery cables to bolt-type, 69
toilets, 114–15
tools (power), operating with battery power, 105–9
transformers, 12, 13–15, 17, 40, 173–74
truck (as a source of backup power), 50, 134–37, 148, 150
truck camper(s), 72, 115, 138, 139–41, 143, 150, 166
TV and DVD players, 95–98
two-meter band radios, 86–88
2 Percent Wire Loss Table, 188

U

Ukraine, cyberattack on the power grid in, 13
understanding
    solar system components, 52–70
    watts, amps, and volts, 190–92
Underwriters Laboratories (UL), 56
uninterruptable power supply. *See* UPS devices
United Motors Car Company, 27
USB outlets, 152

useful tables and wiring diagrams, 185–89
utility grid attacks, 13–17
utility lines, 10, 17
UPS devices, systems, 102–3

V

vehicles
    best bugout, 139
    using to obtain battery power, 134–37
ventilator unit, 100–3
virus (computer), 12, 13–14
Volcano Grill/Stove, 159
volts, understanding watts, amps, and, 190–92

W–Z

walkie-talkies, 82–83
washing without grid power, 161–66
water
    hazards of drinking river and lake, 12
    least-expensive way to heat, 160
    using battery power to provide clean, 110–17
    well, 111–14
water filters, 115–17
watts, amps, and volts, understanding, 190–92
weather radios, recommended, 131
well water, 111–14
wire
    and fuses, 67–70
    inverter wire sizing, 195–98
    loss table (2 percent), 187
    maximum wire amperage recommendations (table 3), 189
    sizing solar array, 193–94

Printed in the USA
CPSIA information can be obtained
at www.ICGtesting.com
LVHW010210260923
759280LV00023B/830